Maximum Success with Nemonik Thinking

Obtain what you seek and escape what you suffer.

Dr Auke Schade

Copyright

First Edition
Published 25 February 2022
Dunedin, New Zealand
@ nemonik-thinking.org
ISBN 978-0-473-62246-6

Abstract

Nemonik Thinking will maximize your success with a minimum of effort and resources. Success is to obtain what you seek and escape what you suffer. Manmade extinction problems, such as nuclear war and climate change, are evidence that humanity's way of Conventional Thinking is failing. Conventional Thinking is a poorly understood, corrupted, and outdated set of cognitive tools that maintains static thinking in a hyperdynamic reality. Unfortunately, that is what they taught you at school. Hence, you will need this handbook, because that will make you the best thinker you can be. Nemonik thinking is for the brain what Windows is for the computer. It is a unique cognitive operating system that comprises dynamic thinking and mind management. Dynamic thinking maximizes your external success with an exhaustive checklist of 21 keywords that activate a set of efficient problem-solving tools. On the other hand, mind management maximizes your internal success with concentrated-relaxation and mantras. Nemonik Thinking is as objective as logic and algebra, but a lot easier to learn and apply. It offers the only realistic protection for your life, family, and assets within our hyperdynamic reality. Nemonik Thinking sounds like magic, because it is.

Contents

1.0 INTRODUCTION

What is in it for you?

Why would you even bother to read this book? Miles of library shelfs are already bending under the colossal weight of books promising instant gratification and happiness. Those books are like ambulances waiting at the bottom of the cliff. If you need them, then you have already hit the rocks. Instead, this book prevents problems. Some people might object that it takes too much time to study this book. However, compare that little amount of time with the many years that you have already spent on your education. This handbook will help you to apply that precious education more efficiently. You might be the best thinker in the world, but only this handbook can make you the best thinker you can be. It will maximize your success by turning weaknesses into strengths, and threats into opportunities, with a minimum of effort and resources.

"Success is to obtain what you seek
and escape what you suffer" [1] §62.

Prevention is the best cure. If the rudder does not control the ship, then the rocks will. Either you spend time preventing problems or you will be forced to cure their consequences.

Incompetent thinkers do not learn from their mistakes and therefore they are doomed to repeat them forever. Competent thinkers learn from their mistakes and never make the same one again. However, nemonik thinkers prevent mistakes by learning from the mistakes of others. In accord, this handbook will increase your mindpower and make you the best nemonik thinker you can be.

Many ancient meta-thinkers have thought about thinking. We have to use their accumulated knowledge, because our lives are far too short to reinvent that cognitive wheel. Therefore, I have synthesized 2,500 years of their ideas into Nemonik Thinking, which incorporates all previously proposed ways of thinking.

It would be fair to ask whether the ancient meta-thinkers could tell you anything useful about your technological reality. However, despite all our technological progress, Lao Zi's principle of cosmic balance has not changed. As our modern equal sign (=) shows, we are still using the principle of balance in all our sciences. Without that equal sign there would be no science at all.

Still some doubts? Remember that without Aristotle's logic, you would not even have a computer, internet, or mobile. He was the first programmer writing code for the natural supercomputers with his 'if-then' statements. Modern science and technology have just added the CPU and hard-drive to run his code on artificial computers. Hence, the ideas of the ancient meta-thinkers are still crucial today.

Those who want to survive tomorrow,
have to learn from those who survived yesterday.

You were born an extraordinary survivor. This handbook will help you to activate those natural talents to maximize your success. They will release your hidden genius and make you the best thinker that you can be. You will become able to prevent, rather than cure problems. You will learn to turn weaknesses into strengths, and threats into opportunities. Nemonik Thinking is the only realistic option to protect your life, family, and assets in our hyperdynamic reality.

Learning and applying Nemonik Thinking is easy. If even I can do it, then certainly you can. After all, you have only to memorize 21 simple words. That is nothing compared to the language that you have learned as a child. You do not even have to start from scratch. You already know those words. You only have to learn a new application for them. You have nothing to lose and everything to win. Just keep going one step at the time. Nemonik Thinking only sounds like magic, because it is super magic.

The Problem

You have the most sophisticated
organic computer in the universe
without a user's manual.

A problem is anything that inhibits the maximization of your success. The ownership, solvability, and priority of each problem have to be examined before trying to solve that problem.

Is the problem preventing the maximization of your personal success? If not, then you do not own the problem. Ignore it. Do not waste your time and resources on chasing windmills like Don Quixote. Nevertheless, you might still be a co-owner. For example, the problem might affect a loved one or your collective. Therefore, you might want to solve it.

Many problems are like monkeys. They sit on the shoulders of their owners and jump on the shoulders of the unaware.

That happens especially to inexperienced managers and caretakers. Their charges complain to them about their monkeys and then the monkeys jump. Delegating responsibility and authority will prevent a monkey zoo on your shoulders.

Problems could be categorized as solved, unsolved, and unsolvable. Some problems are repetitive and have already been solved in the past. It is a waste of time and resources to reinvent the wheel. Make your life easy. If you have already solved the problem in the past, then recall the solution from your memory. Then there are the unsolved problems that are solvable. Those problems should be subjected to Nemonik Thinking. Surprisingly, many people try to solve unsolvable problems. That is a waste of your precious time and resources. Avoid the unsolvable problems or learn to live with them. They might require coping tools such as the ones mentioned in Chapter 5.

Use memory to deal with the already solved.
Use Nemonik Thinking to deal with the unsolved.
Use coping tools to deal with the unsolvable.

The Ancients

For millennia, people lived in a static reality, which was natural, stable, and predictable. The seasons and events were cyclic and repeated over and over again. Problems did not change much. In such a reality, memories of the past are reliable predictors of the future. Memory is about storing information in order to move it across space and time. On the other hand, thinking is about processing information in order to generate new information. Therefore, memory, rather than thinking, is the prime cognitive process in a static reality.

If you knew where and how to plant and hunt, you never had to change that knowledge. Changing a successful solution was counterproductive. Mental flexibility was a weakness. That static reality conditioned humans to become static thinkers. That way was so powerful that humans became the apex-predators driving other species towards extinction.

Static thinkers rely predominantly on frozen memories, rather than fluid thinking. Therefore, they are forced to adjust the situation to their fixed way of thinking. They have the impossible task to adjust the wind to their sails. Therefore, in a hyperdynamic reality, static thinkers waste precious resources on control, conflict, detachment, opposition, and win-lose strategies.

Two and halve thousand years ago, the Greek philosopher Aristotle developed logic. More than a thousand years later, that cognitive tool was rediscovered during Europe's Renaissance, which drove the advantageous industrial, biotechnical, and informational revolutions. Those revolutions turned the static reality into our hyperdynamic one. To date, everything changes all the time and reality is in a state of constant flux.

Systemic Cognitive Failure

In our hyperdynamic reality, the past does not predict the future anymore and has therefore become less important. Why would anyone memorize information that is outdated in nanoseconds? Just Google it. However, groupthink and cognitive dissonance maintained the old ways. Cognitive dissonance is the rejection of correct information, which protects incorrect subconscious information by evoking individual fear, frustration, anger, and aggression. Groupthink is a collective way of cognitive dissonance. Therefore, our cognitive skills developed slower than our technology. People remained static thinkers in their new hyperdynamic reality. As a result, humanity is today suffering from *'Systemic Cognitive Failure'*, which is a failing way of static thinking in a hyperdynamic reality.

Systemic cognitive failure has created a *'Triangle of Doom'*. Education fosters science; science fosters technology; and technology fosters *'Manmade Extinction Problems'*. Hence, manmade extinction problems such as climate change, nuclear war, dwindling resources, overpopulation, industrial pollution, extinction of species, and the overall destruction of nature are caused by humanity's systemic cognitive failure [1] [2] [3] [4].

"The more people know,
the stranger the things they begin to develop" [1] §57.

Education Kills Humanity

Society requires an efficient system to select the best candidates for the limited number of positions in its socio-economic hierarchy. Therefore, the educational system nurtures the grading of the competition between students. In addition, the quality of the students memories are much easier to grade than the quality of their thinking. As a result, education has become a competitive-memory-based grading system. That system reinforces its way of thinking with stimuli including praises, prizes, ceremonies, degrees, and ultimately jobs. The punishment for failing students is a dreadful life in the gutters of society.

Memory is static, because it is always about the past. By definition, it cannot change without losing the truth. On the other hand, in a hyperdynamic reality, everything is changing. Such flux requires a high quality of thinking that constantly updates and creates information. Therefore, the educational grading of static memory is counterproductive in a hyperdynamic reality.

Educational competition has counterproductive cognitive effects on the students. It forces them to focus on winning, rather than success. Winning is defeating opponents in competition. On the other hand; *"Success is to obtain what you seek and escape what you suffer"* [1] §62. Hence, incompetent thinkers might win every competition, but are never successful. Instead, nemonik thinkers might never win, but could be very successful.

In the educational system, winning is more important than the truth. In order to win the educational debates, students are forced to replace Socrates' truth with Protagoras' rhetoric. They have to replace critical thinking with criticism; rational thinking with rationalizations; and justice with justifications. Therefore, cheating is prevalent in almost every aspect of our society from sports to business. Cheaters have become winners

"Sages do not strive.
Therefore, no one can strive with them" [1] §22.

More of the same education is not the solution, because reality shows that the educational way of *'Conventional*

Thinking' is leading to self-destruction. Each solution creates more problems driving us deeper into that trap. In accord, Albert Einstein warned us that we cannot solve problems with the same way of thinking that has created them. The growing pressure of manmade extinction problems suggests that the educational way of thinking is failing dramatically. Therefore, it needs a complete overhaul [2].

Humanity's systemic cognitive failure could destroy the entire planet and drive us towards extinction. We are on the sinking Titanic. However, we still have choices. Some escape reality by submerging in alcohol, drugs, or entertainment. They join the musicians on the aft-deck. Or they still try to make more money by selling the lifeboats of their children. Others reduce the competition for survival by locking the exits from the lower decks. Most just join the panic. They just shout and jump up and down.

Unfortunately, no one can escape reality. Often, the escape attempt is counterproductive, because the situation deteriorates when unattended . The *'day after'* the escape, is always worse. That is the hangover time. Despite the escape attempt, you still have to deal with reality. Therefore, true survivors take productive action and plug the leaks in the hull. They start pumping.

Teachings are only valuable if the student is ready.

The affluence that is created by Conventional Thinking weaves an illusionary cocoon of safety around society. Nevertheless, that way of thinking is part of the problem, rather than the solution. Ultimately, it will fail, because it is a static way of thinking in a hyperdynamic reality. Therefore, it creates personal and manmade extinction problems that it cannot solve. If you want to survive the growing pressure, then you have to improve your thinking. The hyperdynamic reality requires the creativity and intuition of your genius in addition to Aristotle's logic and reason. Adopting Nemonik Thinking is the only realistic option to protect your life, family, and assets.

"Only those who are different can be great.
If they were similar, then they would be insignificant" [1] §67.

Nemonik Thinking is important, because it is different. If it was not different from the failing Conventional Thinking, then it would be unimportant. Therefore, you are likely to face an internal conflict between your conscious manager and you subconscious genius.

Your rational manager cannot ignore the external manmade extinction problems. Those problems are a direct and immediate threat to your life, family, and assets. On the other hand, your genius rules your internal reality and is likely to protect its way of Conventional Thinking against the transformation to Nemonik Thinking.

Your genius has many cognitive tools to its disposal to deter your conscious manager from any mental improvement. It could strengthen its firewall with a rhetoric of criticism, rationalizations, and justifications. It could distract your manager by changing the topic over and over again. It could make an infinite number of emotional statements and ignore all rational comments. Or use groupthink and cognitive dissonance to bombard your conscious manager with worrying, stress, fear, frustration, headaches, boredom, and anger.

There is no sense in sugarcoating your personal problems. No one promised you a rose garden. Some people lose their lives, because they ignore the external reality. Others lose their lives, because they ignore the internal reality. You have to take care of both. Indeed, there are thousands of reasons to defend Conventional Thinking. On the other hand, there are only three reasons to accept Nemonik Thinking. Those are your life, family, and assets. However, that are the only reasons that really count.

Maximizing your success requires your conscious to become a competent manager of your subconscious genius. Your rational manager knows that manmade extinction problems are scientific evidence for humanity's systemic cognitive failure. That failure includes the conventional way of thinking that you have learned at school. The internal and external pressures will go through the roof. Just look at the global havoc caused by the Covid pandemic. If you still have any doubts, then consider the fate of the millions of already displaced people. Even surviving the fallout of one manmade extinction problem will require all the mindpower you can obtain.

Unfortunately, the external reality could not care less what you decide. Furthermore, arguing with me and rejecting my ideas is a waste of your precious time. Nemonik Thinking is not a religion. I am not a guru or a priest. It is not my mission to convert you or save your afterlife. I do not get rewards for the number of converts. I have no superpowers either. I cannot even start to solve your problems. I can only create cognitive tools so that you can solve your own. My mission is giving people the opportunity to become the best thinkers they can be.

It is unimportant whether I can convince you to use Nemonik Thinking. Instead, it is crucial whether you can convince yourself, because your life, family, and assets are on the line. It is up to you to break the subconscious mental shackles that imprison your conscious. That fight is inside your mind, rather than outside. Success!

The Solution

Never doubt your natural talents, because you are a born survivor and apex-predator. You are the descendant of an unbroken chain comprising thousands of survivors. Your ancestors lived in caves, fought dangerous bears with fragile flint knives, and hunted colossal mammoths with wooden spears. They survived disasters such as wars, earthquakes, diseases, famines, floods, tornados, and ice ages. Otherwise, you would not be here.

Show me a person without problems
and I show you a dead one.

You are unique, because no one else has your genes. Nature has honed your genes to perfection for millennia. You are nature's optimal solution for whatever problems you have to face. Relax! Problems are no exceptions. Life is a chain of problems. You only have to discover your natural talents for problem-solving. This handbook provides you with a shortcut.

The Mind

The mind is the nonmaterial part of a person that includes all conscious, semiconscious, and subconscious processes and structures resulting from interactions between the brainware and mindware of that person. Brainware is a set of material components comprising the Central Nerve System. It could be compared to the hardware of a computer. In contrast, mindware is a set of nonmaterial procedures that could be compared to the software of a computer.

The mind is inevitably associated with knowledge. Knowledge is a description of reality, while wisdom is the productive application of that knowledge. On the other hand, ignorance is a lack of knowledge.

The two cognitive procedures to obtain knowledge are memory and thinking. Memory is about storing knowledge in order to move it across space and time. On the other hand, thinking is about processing knowledge in order to generate new knowledge. Memory could be compared to the hard-drive of a computer, while thinking could be compared to its CPU.

The quality of your mind
determines the quality of your life.

The mind has to store, retrieve, and process a gigantic amount of external and internal data. If a person would be consciously aware of those cognitive processes, then that person would suffer from an information-overload that would inhibit proper functioning. Therefore, the mind is naturally divided in a conscious and subconscious. Hence, there is nothing supernatural about the existence of a hidden subconscious.

Conscious

As illustrated in the following Figure 1, the conscious is the relatively small part of the mind that is dominant when awake; aware of the reality and the Self; manages the mind; and maximizes success with Rational Thinking.

Rational Thinking is a conscious way of thinking that subjects valid and reliable descriptions of reality to logic and reason to create new valid and reliable descriptions of reality. Rational Thinking comprises the OBJECTIVE and COLLECTIVE mindmodes. A mindmode is a special way of thinking that deals with a particular topic. The OBJECTIVE mindmode deals with the laws of nature, while the COLLECTIVE mindmode deals with the rules of people.

The subconscious is the largest part of the mind that is dominant when asleep; stores and processes bulk information; is consciously inaccessible; and maximizes success with Affectorial Thinking.

Affectorial Thinking is a subconscious way of thinking that modifies and creates mindsets. A mindset is a habituated set of rules or algorithms that generates affecters. An affecter is a mental signal that influences the conscious.

Affectorial Thinking comprises the CREATIVE and REACTIVE mindmodes. The CREATIVE mindmode deals with the future by generating mindsets that produce affecters such as: ideas, innovations, and intuitions. The REACTIVE mindmode deals with the past by generating mindsets that produce affecters such as: skills, emotions, and feelings.

<table>
<tr><th colspan="5">MIND</th></tr>
<tr><th colspan="2">CONSCIOUS</th><th>SEMI CONSCIOUS</th><th colspan="2">SUB CONSCIOUS</th></tr>
<tr><td colspan="2">Manager
Awake
Rational-thinking</td><td>Communicator
Drowsy
Concentrated-relaxation</td><td colspan="2">Genius
Asleep
Affectorial-thinking</td></tr>
<tr><td>nature</td><td>people</td><td rowspan="2">Communication channel</td><td>future</td><td>past</td></tr>
<tr><td>OBJECTIVE mindmode</td><td>COLLECTIVE mindmode</td><td>CREATIVE mindmode</td><td>REACTIVE mindmode</td></tr>
<tr><td colspan="2">mantras ></td><td>< firewall ></td><td colspan="2">< affecters</td></tr>
</table>

Figure 1: Model of the mind.

Subconscious

Most of the cognitive information is processed and stored in the larger subconscious. Hence, that protection against information overload creates a subconscious genius. That genius stores and processes information and executes all inherited and learned skills. That includes walking, talking, driving the car, mathematics, playing chess and music, karate, etc.

The genius has to protect itself from external and internal interference. Therefore, it treats contradictory information as a malignant cognitive virus. To stop such viruses, the genius blocks them with a *Mental Firewall.* That firewall is a cognitive tool that protects the genius from external and internal interference with cognitive dissonance and groupthink.

Conscious dominance is a healthy mental state of being awake that is fostered by concentration, while subconscious dominance is a healthy state of being asleep that is fostered by relaxation. Conscious and subconscious dominance are mutually exclusive. As a result, conscious Rational Thinking is also mutually exclusive from subconscious Affectorial Thinking. The logic and reason of the head inhibit the skills, emotions, and feelings of the heart and vice versa.

The heart knows reason
that reason does not understand.

You need all your accumulated logic, reason, skills, emotions, and feelings to maximize your success in our hyperdynamic reality. If you would even ignore one of the four mindmodes, then you would be a static thinker in a dynamic universe. No matter how successful, such thinkers fail when the situation inevitably changes.

Even when fully awake, the subconscious could become dominant during an emergency. For example, while driving your car, you might hit the brakes without any conscious thought. You just hit the brakes with a *'silent mind'.* A silent mind is a mental state that activates the subconscious without conscious interference, which is often used to cope with emergencies. Also referred to as the Empty Mind or Zone.

Semiconscious

The semiconscious is the communication channel between the conscious and subconscious. Semiconscious dominance is a drowsy healthy state of mind that lowers the firewall between the conscious manager and the subconscious genius. Every time you fall asleep or wake-up, you slide automatically through that mental twilight state.

Nemonik Thinking

Nemonik Thinking is for the brain what Windows is for the computer. Nemonik Thinking is a unique cognitive operating system. It maximizes your success within the external reality with Dynamic Thinking (Chapter 2); and within the internal reality with Mind Management (Chapter 3) [1] [5] [6]. *"Success is to obtain what you seek and escape what you suffer"* [1] §62. The external reality surrounds your mind and includes space, matter, and time. On the other hand, the internal reality is the mental perception of that external reality. Nemonik Thinking will make you the best thinker you can be by releasing your hidden genius.

Nemonik Thinking is based on: the cosmic principle of balance [7]; the first practical model of the mind as presented in the preceding Figure 1; natural cognitive association; and the time, matter, and space dimensions of modern physics [6]. Nemonik Thinking integrates ideas of meta-thinkers such as Aristotle, Confucius, Edward de Bono, Lao Zi, Plato, Socrates, and Sun Zi. Therefore, it is a framework for all previously proposed cognitive tools such as: analytical, collective, creative, critical, intuitive-emotive, lateral, left-brain, legalistic, logical, objective, rational, reactive, right-brain, scientific, strategic, and systems thinking.

Nemonik Thinking does not promote any mysticism, cult, or ideology. The supernatural is outside the scope of Nemonik Thinking. It does not try to change what you think or believe. It does not replace anything. Instead, it aims to improve the efficiency of your thinking and to optimize the application of your personal experience, skills, knowledge, and wisdom. It will use every bit of information that you have already collected. It is an informational coat-hanger that organizes information more efficiently. Nemonik Thinking offers a friendly makeover, rather than a hostile takeover. It provides cognitive tools that are as objective as logic and algebra, but a lot easier to learn. After all, you have only to memorize 21 simple words.

Dynamic Thinking

The educational way of Conventional Thinking fosters humanity's systemic cognitive failure. It is a competitive-memory based way of conscious thinking that aims to win competitions in the external reality by using free-association.

Free-association is a cognitive procedure that activates spontaneously a link between two or more pieces of information. For example, imagine consciously the word *'black'*. That might bring automatically associated words to mind such as dark, shadow, night, and Yin. But also, opposites such as white, day, light, and Yang. As a result of the unsystematic free-association, conventional thinkers cannot predict consciously what their search might return. Their minds are like libraries filled with unsorted books. To find a particular book, they might have to search their entire collection. As Murphy's law suggests—*It is always the last one.*

Dynamic Thinking is the part of Nemonik Thinking that maximizes external success by using the *Nemonik Associative Network* (Network) and guided-association.

As illustrated in the following Figure 2, the Network is a subconscious construction linking an exhaustive set of 21 keywords called nemoniks with guided-association to an almost infinite amount of well organized problem-solving tools. Guided-association is a cognitive procedure that activates a memorized link between a nemonik and the icon, definition, potential questions, SWOT analyses, mantras, and proverbs for that nemonik.

The following Figure 2 presents the hierarchy of the *Nemonik Associative Network* (Network) that comprises 21 nemoniks. A nemonik is a memory and thinking tool that recalls associated information from memory and activates the processing of information to create new information.

The 21 nemoniks form a kind of mental keyboard. Each key provides an access code to the specific problem-solving tools for that nemonik. That imaginary keyboard reduces significantly the mental recalling and processing time of information, because it transforms the internal chaos into order.

The Oxford Dictionary contains about 170,000 words. Imagine, that those words were printed in a random order. How long would it take you to find a particular word in that chaos? You might have to read the entire dictionary. Fortunately, the editors have sorted the words in alphabetical order. Therefore, you have only to search less than 4% of the words.[1] As a result, it will take only a few seconds to locate any word in that dictionary.

NEMONIK ASSOCIATIVE NETWORK	Principle	Lao Zi's Cosmic Balance	**Focus** **Balance** **Efficient** **Proactive**
	Mind	Conscious	**Objective** **Collective**
		Sub-conscious	**Creative** **Reactive**
	Reality	Space	**Advance** **Stay** **Retreat**
		Matter	**Accumulate** **Preserve** **Dispose**
		Time	**Act** **Wait** **Prepare**
	Interaction	Perception	**Accept** **Reject**
		Projection	**Reveal** **Conceal**

Figure 2: Nemonik Associative Network

The same is true for a library where books are randomly placed on shelves. Some libraries have more than a million books. How long would it take you to find a particular book

[1] 100% / 26 letters = 3.8%

in such a chaotic library? Therefore, libraries organize the books by topic with the Dewey decimal number system. Hence, access codes foster the efficient recall and processing of information.

Compared to a dictionary or library, the brain is a huge organic computer with billions of neurons each one interconnected with multiple synapses. Unfortunately, it has no handbook or keyboard. As a result, conventional thinkers do not have a conscious organized cognitive system or efficient access codes for their computer. Their mind is like a random dictionary or library. Therefore, their recall of information is inefficient, inaccurate, and slow.

How much processing time is wasted by conventional thinkers to find the specific problem-solving tools in that chaos? How can they be sure that they have retrieved or processed all available information for a particular problem? How do they know what is missing? Therefore, you need the 21 nemoniks.

Nemoniks sort the information in your mind and provide access codes for retrieving and processing that information. Nemoniks transform mental chaos into order, which will save you time and effort.

The mind of a dynamic thinker is like a well-organized library. Each piece of problem-solving information has a nemonik access code and is stored systematically on the imaginary library shelf associated with that nemonik. Nemonik thinkers have only to search less than 5% of their memory in order to recall the associated problem-solving tools including the icon, definition, potential questions, SWOT analyses, mantras, and proverbs for each nemonik.

As shown in the preceding Figure 2, the 21 nemoniks are: Focus, Balance, Efficient, Proactive [7]; Objective, Collective, Creative, Reactive; Advance, Stay, Retreat; Accumulate, Preserve, Dispose; Act, Wait, Prepare; Accept, Reject; Reveal, and Conceal [6].

In Lao Zi's philosophy, balance is the cosmic principle that rules our external reality. All perceived chaos, movement, reaction, and entropy are manifestations of forces restoring that balance. For example, a stone thrown into the water creates waves. Those waves manifest the forces restoring the water balance. No matter how many stones you throw, the water will always return to a balanced and tranquil state. Lao Zi's principle of balance is supported by the equal sign (=) that is crucial in modern mathematics, science, and our thinking.

Forces restoring the balances could distract you from your goal, because moving waves are more spectacular than a tranquil surface. Therefore, it is prudent to maintain a strong FOCUS on your success. *Ask for nothing and you will get nothing.* Any action that disturbs the balance could evoke counterproductive reactions. Therefore, it is productive to maintain the BALANCE. *Do not rock the boat.*

During the restoration of a disturbed balance, it pays off to be EFFICIENT and align with the restoring forces. *Go with the flow.* On the other hand, if an imbalance could create a counterproductive restoration in the future, then it is sensible to be PROACTIVE and restore that balance before it can do any harm [7]. *Prevention is the best cure.*

As shown in the preceding Figures 1 and 2, the conscious is divided in OBJECTIVE and COLLECTIVE mindmodes, while the subconscious is divided in CREATIVE and REACTIVE mindmodes. The OBJECTIVE mindmode deals with the laws of nature, while the COLLECTIVE mindmode deals with the rules of people. Furthermore, the CREATIVE mindmode deals with the future, while the REACTIVE mindmode deals with the past [6].

As shown in the preceding Figure 2, the exhaustive aspects of the external reality are: space, matter, and time. Your spatial options are limited to ADVANCE, STAY, and RETREAT in regard to your goal. Your material options are limited to ACCUMULATE, PRESERVE, and DISPOSE resources. Your temporal options are limited to ACT, WAIT, or PREPARE [6].

As shown in the preceding Figure 2, the interactions between the mind and the external reality comprise the perception and projection of information. The perception of information from reality to the mind provides the options to ACCEPT or REJECT that information. On the other hand, the projection of information from the mind to reality provides the options to REVEAL and CONCEAL that information [6].

Memorizing the set of 21 nemoniks and their associated information will install the Network in your mind (Chapter 2). That network is exhaustive, because it takes into account each aspect of problem-solving. It delivers problem-solving tools and information at your fingertips.

Dynamic thinkers rely predominantly on fluid thinking, rather than frozen memories. There are no good or bad nemoniks. The value of each nemonik depends entirely on the actual situation. For each problem, dynamic thinkers will recall and evaluate each of the 21 nemoniks. They maximize their success by continuously adjusting their thinking to the hyperdynamic reality. Like sailors, adjusting their sails to the wind.

Dynamic Thinking is natural, efficient, proactive, and systematic; it prevents stagnation; prepares your mind for problems before they appear; and inhibits cognitive blind spots. Dynamic thinkers are goal-oriented, rather than competition-oriented. Therefore, they foster freedom, alignment, compassion, allies, and win-win strategies. Dynamic Thinking is easy to learn and simple to apply. The exercises are described in Chapter 2.

Mind Management

Mind Management is the part of Nemonik Thinking that maximizes internal success by using Concentrated-Relaxation and Mantras. The aim of Mind Management is fostering conscious management of the subconscious genius by opening their semiconscious communication channel.

Virtual reality

Sometimes, people claim that they know reality: *"I saw it with my own eyes".* However, what did they see? What is that reality? Reality includes the external and internal realities. The external reality surrounds your mind and includes space, matter, and time. In contrast, the internal reality is your mental perception of that external reality.

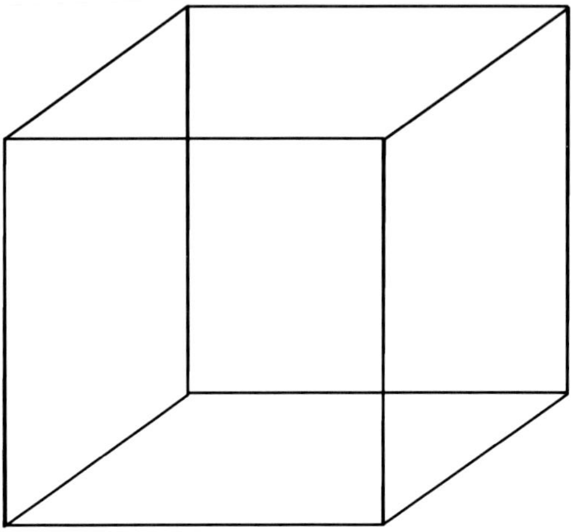

Figure 3: Necker cube.

Figure 3 shows the optical illusion that is referred to as the *Necker Cube.* If you look carefully at the cube, then the front keeps flipping back and forwards. Your subconscious tries to interpret the ambiguous 2-D image as a true 3-D one. However, the Necker cube provides two correct sensory solutions. Therefore, the subconscious cannot stabilize the picture. It presents alternatively one of the two 3-D options to your conscious.

Although your internal reality is normally seamless, the Necker cube shows a glitch in that reality like a defect TV screen. Suddenly, in the middle of a movie, the illusion is broken. You see the real flickering screen, rather than the illusion created by the video. The Necker cube shows you a window into the true external reality. The normally stable reality flickers between two incompatible pictures. That optical illusion shows that your subconscious is trying to create an illusionary but stable internal reality. That supports the notion that your subconscious genius determines your conscious perception and experience of the external reality.

As they say: *There is nothing new under the Sun.* Modern computers are not the first or only ones that create virtual realities. As shown in the following Figure 4, your conscious lives in a virtual reality that is created and maintained by your subconscious genius. Your conscious is caught within that reality.

The idea that you live in a virtual reality is supported by your illusionary perceptions. Physics shows that there is only energy and matter in the external reality, which consist of frequencies, wavelengths, and atoms. There are no phenomena such as colours, sounds, odours, and tastes in that reality. Consequently, your genius has to translate the external energy and matter into internal and subjective perceptions.

"My words are very easy to understand
and very easy to apply.
Yet, people cannot understand them
and they cannot apply them" [1] §70.

It is hard to believe, but everything that you experience is created by you. Hence, that happy sound of a child, the scent of a rose, and the taste of your favourite food are created by you. Even the beautiful rainbows and sunsets that you have seen are virtual illusions created by your subconscious for your conscious benefit.

If a tree falls in a forest,
and there is no one there,
would there be a sound?

Physics reveals also that the solid matter you perceive in your virtual reality is about 99% space in the external reality. Hence, there is not much matter in solid matter. An x-ray provides a far better representation of solid matter than your visual perception. Moreover, Albert Einstein explained that time, which is so accurately measured by our clocks, is in fact relative, rather than constant. Hence, you live in an unreal reality.

Feelings such as fear, pain, happiness, joy, and sadness are mental signals produced by your genius to affect your conscious. Those feelings do not exist outside the human mind. They are part of your internal virtual reality, rather than the external one. If your feelings hurt you, then do not try to solve them in the external reality.

"Why do I have great suffering?
That is, because I have a body.
If I had no body, how could I suffer?" [1] §13.

It might be disappointing, but you live in a relative, black-and-white, silent, tasteless, uncompassionate, and transparent ghost world. I know, that is hard to believe. However, imagine how fantastic that is. Your genius transforms that bleak external reality into the fantastic virtual reality you live in. Wherever you are and whatever you do; you find yourself always at the centre of that virtual reality. That illusion is the true miracle of life.

Your genius has enormous powers over your conscious experience of life. It can turn the external reality from heaven into hell, and vice versa. The richest person on this planet could be still unhappy, while the poorest might be jubilant and joyful. The only one that could cause you pain is you. Hence, all your efforts to maximize your external success could be in vain if you do not maximize your internal success. Fortunately, you are omnipotent in your virtual reality, because you create and maintain it yourself. Therefore, you should take extreme care to manage consciously your subconscious genius with the tools of Mind Management.

You can only manage your feelings
after you have acknowledged that you create them.

Mind Management

As shown in the following Figure 4, Nemonik Thinking comprises Dynamic Thinking and Mind management. Dynamic Thinking uses the Nemonik Associative Network of 21 nemoniks to maximizes your external success. On the other hand, Mind Management uses Concentrated-Relaxation and Mantras to maximizes your internal success.

Concentrated-Relaxation is a cognitive tool that evokes semiconscious dominance by reducing conscious dominance without falling asleep. In turn, semiconscious dominance lowers the protective firewall, which opens the communication channel between the conscious manager and the subconscious genius. This allows mantras to affect the subconscious. A mantra is a word or phrase that is repeated in the mind during semiconscious dominance in order to create mindsets. Mindsets are habituated set of rules or algorithms that generate affecters. Affecters are mental signals generated by a mindset that influences the conscious.

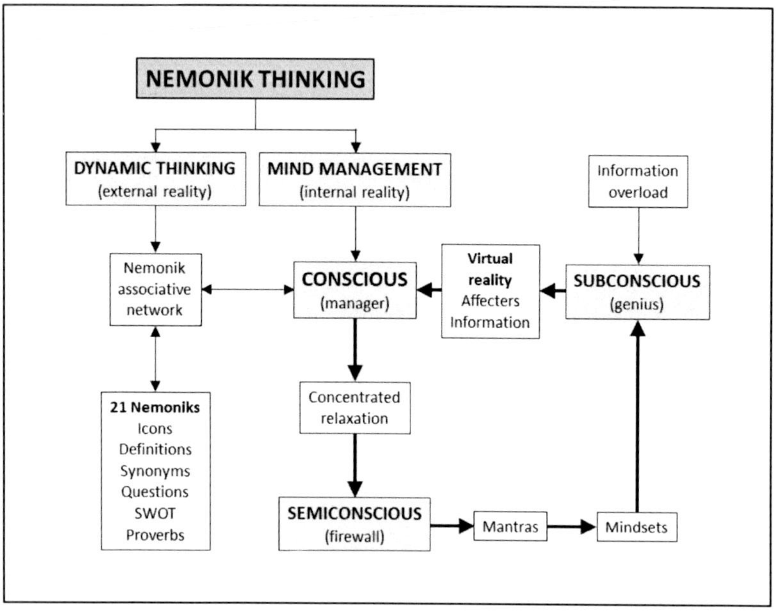

Figure 4: Flowchart Nemonik Thinking.

Mantras strengthen the associative links of the Network and foster balances, within and between, the external and internal realities. Ultimately, those balances create the elusive peace of mind and enlightenment.

Summary

The snowballing manmade extinction problems suggest that humanity suffers from systemic cognitive failure. Hence, there is not enough time left to develop a valid and reliable alternative for Nemonik Thinking [3] [4]. I should know, because it took me a lifetime to develop this unique cognitive tool. It might not be perfect, but it is the only available one.

"If competent people hear about the Way,
then they are able to practise it constantly.
If mediocre people hear about the Way,
then they put it in a safe place and seem to lose it.
If incompetent people hear about the Way,
then they laugh loudly about it" [1] §41.

Systemic cognitive failure and manmade extinction problems are facts of life. Of course, Nemonik Thinking is different. Otherwise, it would not help. Humanity is running out of time. Procrastination and using rhetoric to argue for arguments sake will not help you. Reality does not care about criticism, rationalizations, and justifications maintaining the status quo. If you are not part of the solution, then you are part of the problem. Are you ready to release your genius with Nemonik Thinking? This might be your last chance to protect your life, family, and assets in this hyperdynamic reality.

The teacher will come when the student is ready.

This handbook provides a unique foundation for the fields of education and psychology. It is a guideline for the survivors aiming to maximize their success by synthesizing their cognitive processes with the hyperdynamic external reality. It restores the external and internal balances. It presents an urgent warning for the teachers that they have to upgrade their failing way of thinking. It is a lifeline for the next generation. This handbook provides a crucial ideology for the advocates of peace and the environmentalists that is firmly based on physics. It is written for you and your loved ones.

2.0 EXERCISES DYNAMIC THINKING

Dynamic Thinking is the part of Nemonik Thinking that maximizes external success by using the *Nemonik Associative Network* (Network) and guided-association. The Network is a subconscious construction linking an exhaustive set of 21 keywords called nemoniks with guided-association to an almost infinite amount of well organized problem-solving tools. Guided-association is a cognitive procedure that activates a memorized link between a nemonik and the icon, definition, potential questions, SWOT analyses, mantras, and proverbs for that nemonik.

2.1 Nemonik Cards

You are invited to make a deck of nemonik cards. As illustrated in the following Figure 5, the easiest way is to select 21 cards from a deck of normal playing cards. Alternatively, you could also use business cards etc. Write the name of a different nemonik on each card.

Focus, Balance, Efficient, Proactive, Objective, Collective, Creative, Reactive; Advance, Stay, Retreat; Accumulate, Preserve, Dispose; Act, Wait, Prepare, Accept, Reject, Reveal, and Conceal.

The cards could help you with memorizing the nemoniks and evaluating your knowledge of their associated icons, definitions, synonyms, SWOT analyses, questions, and proverbs. They will also help you to create your nemonik profile.

Figure 5: The 21 nemonik cards.

2.2 Nemoniks and Icons

The aim of this exercise is memorizing the exhaustive set of 21 nemoniks and their associated icons as presented in the following Table 1. A nemonik is a memory and thinking tool that recalls associated information from memory and activates the processing of information to create new information. There are no good or bad nemoniks. The value of each nemonik depends entirely on the actual situation.

The set of 21 nemoniks is the basis for your subconscious *Nemonik Associative Network* (Network). As illustrated in the following Figure 6, each nemonik is a conscious access code to a specific subconscious location in your Network. That fosters the efficient recall and processing of information associated with each nemonik. For example, consciously recalling the access code [ADVANCE] is like typing an instruction on the keyboard of the imaginary laptop in your conscious. That access code will lead your conscious directly to the correct subconscious location. Your subconscious will then return ADVANCE and its associated icon straight to the imaginary screen of your conscious.

Just memorizing the 21 nemoniks is not good enough. Everyone knows those words. They are common words, rather than magic spells. You have to apply them diligently to every problem, solution, or strategy you encounter. You have to practice them in the same way as karate experts practice their katas over and over again. Just reading a book about karate does not make you a martial expert. You have to repeat those nemoniks consciously until your subconscious has habituated mindsets for them. You have to be able to use them with a silent mind without any conscious thought. It has to be automatic. That is when your genius starts the pure magic.

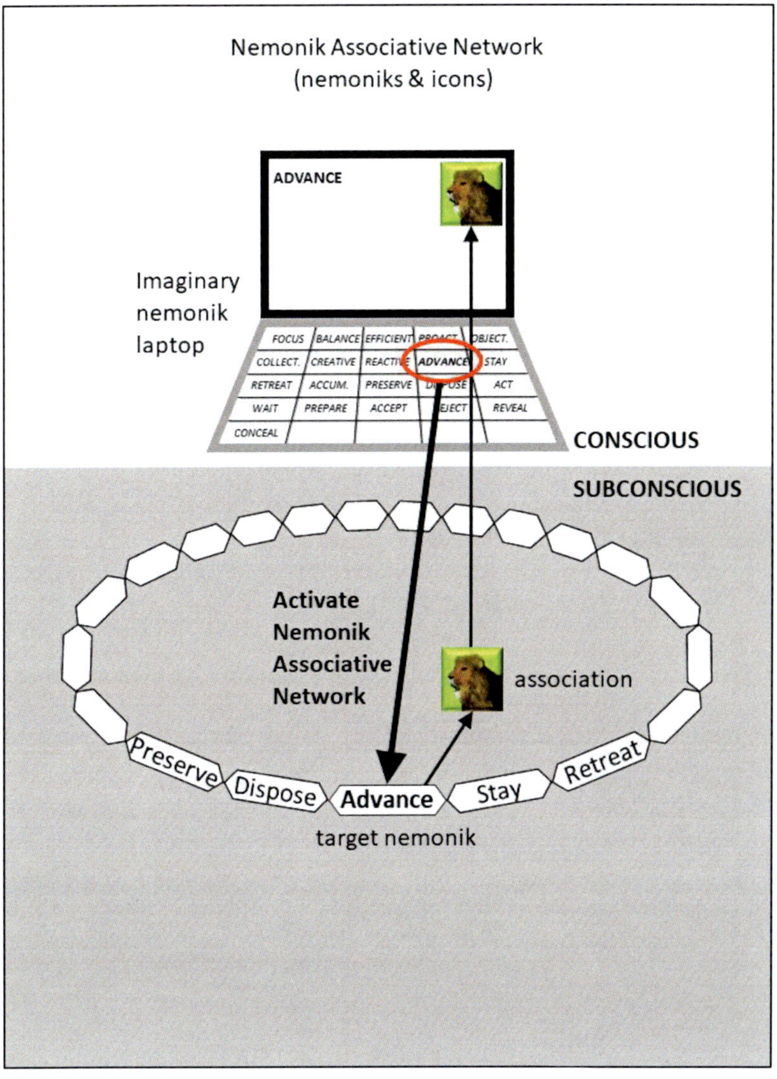

Figure 6: Nemoniks and icons.

Table 1: Study nemoniks and their icons.	
	**FOCUS** *Archers succeed because they focus on their target.*
	**BALANCE** *Traders succeed because they balance their products with standard weights.*
	**EFFICIENT** *Sailors succeed because they are efficient and set their sails to the wind.*
	**PROACTIVE** *Workers succeed because they are proactive and wear helmets to prevent injury.*
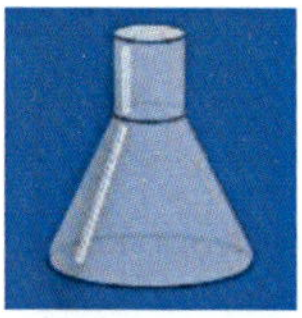	**OBJECTIVE** *Scientists succeed because they use natural laws to predict the future.*
	**COLLECTIVE** *Lawyers succeed because they use artificial rules to predict the future.*

Table 1: Study nemoniks and their icons.	
	**CREATIVE** *Inventors succeed because they use creativity to deal with the future.*
	**REACTIVE** *Programmers succeed because they apply reactive algorithms learned in the past.*
	**ADVANCE** *Lions succeed by advancing with speed on their prey.*
	**STAY** *Rabbits succeed by staying motionless in position until the danger has passed.*
	**RETREAT** *Deer succeed by retreating rapidly from danger.*
	**ACCUMULATE** *Squirrels succeed by accumulating nuts for the winter shortages.*

Table 1: Study nemoniks and their icons.	
	**PRESERVE** *Bees succeed by preserving honey from deterioration.*
	**DISPOSE** *Vultures succeed by disposing the waste of nature.*
	**ACT** *Snakes succeed by acting with total determination at the right time.*
	**WAIT** *Spiders succeed by patiently waiting for their prey to fall in their trap.*
	**PREPARE** *Beavers succeed by preparing dams to raise the water level.*
	**ACCEPT** *Puppies succeed because they accept everything you offer them.*

Table 1: Study nemoniks and their icons.	
	**REJECT** *Hedgehogs succeed because they reject everything you offer them.*
	**REVEAL** *Elephants succeed because they reveal their size and strength as a deterrent.*
	**CONCEAL** *Chameleons succeed because they conceal themselves by adjusting their colour to the environment.*

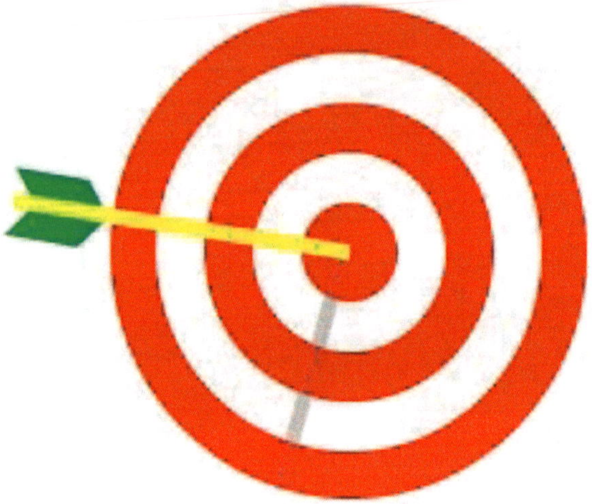

Test Nemoniks and Icons

Instructions
Write the nemonik for each icon and its corresponding number on a piece of paper. The correct answers will be on the page after the questions.

Name the nemonik for each icon.	
	1
	2
	3
	4

Answers: nemonik for each icon.	
1	Proactive.
2	Accumulate.
3	Stay
4	Reject.

Name the nemonik for each icon.	
	**5**
	**6**
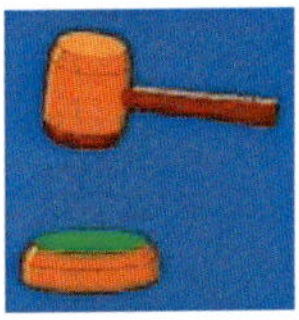	**7**
	**8**
	**9**
	**10**

Answers: nemonik for each icon.	
5	Prepare.
6	Objective.
7	Collective
8	Advance
9	Preserve
10	Accept.

Name the nemonik for each icon.	
	11
	12
	13
	14
	15
	16

Answers: nemonik for each icon.	
11	Retreat.
12	Balance.
13	Creative.
14	Act.
15	Reveal.
16	Focus.

Name the nemonik for each icon.	
	**17**
	**18**
	**19**
	**20**
	**21**

Answers: nemonik for each icon.	
17	Reactive.
18	Wait.
19	Conceal.
20	Dispose.
21	Efficient.

Before moving on to the next exercise, make sure that you can recall the nemonik for each icon and vice versa. The following exercises aim to expand your Network with definitions, synonyms, SWOT analyses, questions, and proverbs.

2.3 Nemoniks and Definitions

The aim of this exercise is to extend the subconscious *Nemonik Associative Network* (Network) by memorizing the definition of each nemonik as described in the following Table 2. Using those definitions avoids confusion and wasting time on recalling and processing ambiguous descriptions. Clear definitions also improve the communication between nemonik thinkers.

As illustrated in the following Figure 7, consciously recalling the access code [ADVANCE DEFINITION] will activate your Network. For nemonik thinkers, that code will return: *"Advance prompts the mind to maximize success by decreasing the distance to the goal"*. Hence, the access code will lead the conscious straight to the subconscious definition and other problem-solving tools associated with that nemonik.

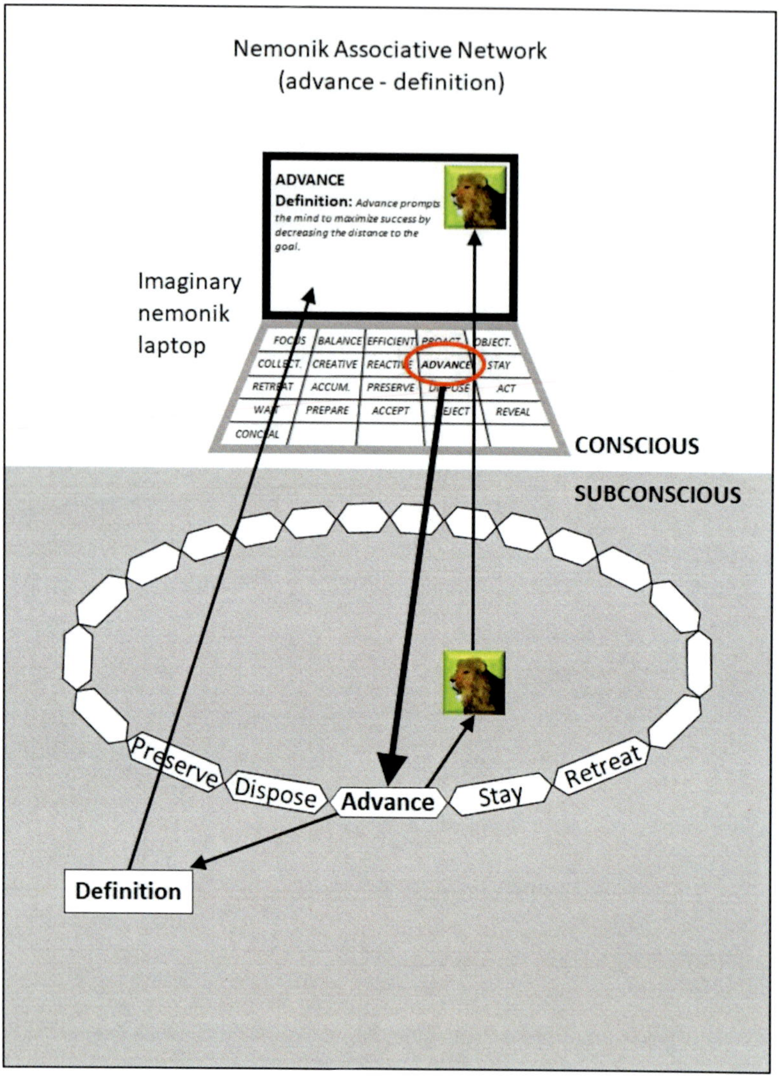

Figure 7: Advance and definition.

Table 2: Study nemoniks and their definitions.

	FOCUS—prompts the mind to concentrate on maximizing success. Success is to obtain what you seek and escape what you suffer.
	BALANCE—prompts the mind to maximize success by maintaining the balances, which prevents counterproductive rebalances.
	EFFICIENT—prompts the mind to maximize success with a minimum of effort and resources by aligning with rebalancing forces.
	PROACTIVE—prompts the mind to maximize success by restoring the balances before the problems grow.
	OBJECTIVE—prompts the conscious mind to maximize success by using Rational Thinking to deal with the laws of nature.
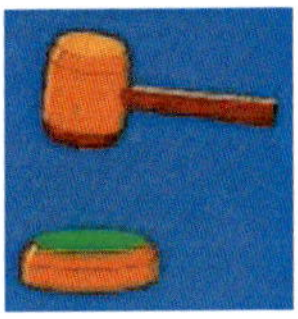	COLLECTIVE—prompts the conscious mind to maximize success by using Rational Thinking to deal with the rules of people.

Table 2: Study nemoniks and their definitions.	
	CREATIVE—prompts the subconscious mind to maximize success by using Affectorial Thinking to deal with the future.
	REACTIVE—prompts the subconscious mind to maximize success by using Affectorial Thinking to deal with the past.
	ADVANCE—prompts the mind to maximize success by decreasing the distance to the goal.
	STAY—prompts the mind to maximize success by maintaining the same distance to the goal.
	RETREAT—prompts the mind to maximize success by increasing the distance to the goal.
	ACCUMULATE—prompts the mind to maximize success by increasing the amount of matter that is under control.

Table 2: Study nemoniks and their definitions.	
	PRESERVE—prompts the mind to maximize success by maintaining the same amount of matter that is under control.
	DISPOSE—prompts the mind to maximize success by decreasing the amount of matter that is under control.
	ACT—prompts the mind to maximize success by changing or moving matter in space and time.
	WAIT—prompts the mind to maximize success by delaying action until it is the most productive time for that action.
	PREPARE—prompts the mind to maximize success by getting ready for the most productive action.
	ACCEPT—prompts the mind to maximize success by accepting the incoming information as a true description of the external reality.

Table 2: Study nemoniks and their definitions.	
	REJECT—prompts the mind to maximize success by rejecting the incoming information as a true description of the external reality.
	REVEAL—prompts the mind to maximize success by projecting true information to the external reality.
	CONCEAL—prompts the mind to maximize success by hiding true information from the external reality.

Test Nemoniks and Definitions

Instructions

Write the nemonik for each definition and its corresponding number on a piece of paper. The correct answers will be on the page after the questions.

Name the nemonik for each definition.	
1	Mindmode that uses Rational Thinking to deal with the rules of people.
2	Mindmode that uses Rational Thinking to deal with the laws of nature.
3	Concentrate on maximizing success.
4	Decide that the incoming information is a true description of the external reality.
5	Change or move matter in space and time.
6	Decrease the amount of matter that is under control.
7	Decrease the distance to the goal.
8	Delay action until the most productive moment for that action.
9	Get ready for the most productive action.
10	Hide true information from the external reality.

Answers: nemonik for each definition.	
1	Collective.
2	Objective.
3	Focus.
4	Accept.
5	Act.
6	Dispose.
7	Advance.
8	Wait.
9	Prepare.
10	Conceal.

Name the nemonik for each definition.	
11	Increase the amount of matter that is under control.
12	Increase the distance to the goal.
13	Maintain the same amount of matter that is under control.
14	Maintain the same distance to the goal.
15	Prevents counterproductive reactions.
16	Project true information to the external reality.
17	Decide that the incoming information is not a true description of the external reality.
18	Restore the balance before the problem grows.
19	Use a minimum of effort and resources to obtain a maximum result.
20	Mindmode that uses Affectorial Thinking to deal with the future.
21	Mindmode that uses Affectorial Thinking to deal with the past.

Answers: nemonik for each definition.	
11	Accumulate.
12	Retreat.
13	Preserve.
14	Stay.
15	Balance.
16	Reveal.
17	Reject.
18	Proactive.
19	Efficient.
20	Creative.
21	Reactive.

Before moving on to the next exercise, make sure that you can recall the definition for each nemonik. The following exercises aim to expand your Network with synonyms, SWOT analyses, questions, and proverbs.

2.4 Nemoniks and Synonyms

The aim of this exercise is extending your subconscious *Nemonik Associative Network* (Network) by studying the synonyms associated with each nemonik as described in the following Table 3. Synonyms are words with a similar meaning as the particular nemonik.

As illustrated in the following Figure 8, replacing the multitude of synonyms with a single nemonik will reduce the mental noise. For nemonik thinkers, the synonyms *'attack, bypass, encroach, forward, infiltrate, invade, offensive, progress, vanguard, etc.'* will all lead their conscious to the nemonik ADVANCE. Although a nemonik could have more than a hundred synonyms, you have only to think about one simple nemonik. Hence, the noise reduction could be more than 99%.[2]

On the other hand, consciously recalling the access code [ADVANCE SYNONYMS] will return: *'attack, bypass, encroach, forward, infiltrate, invade, offensive, progress, vanguard, etc.'* Hence, the access code will lead the conscious straight to the subconscious synonyms and problem-solving tools associated with that nemonik.

Remember, you have to study those synonyms until your subconscious has habituated mindsets for them. That is when your genius starts the pure magic.

---

[2] Noise reduction: ((100 – 1) synonyms / 1 nemonik) = 99%.

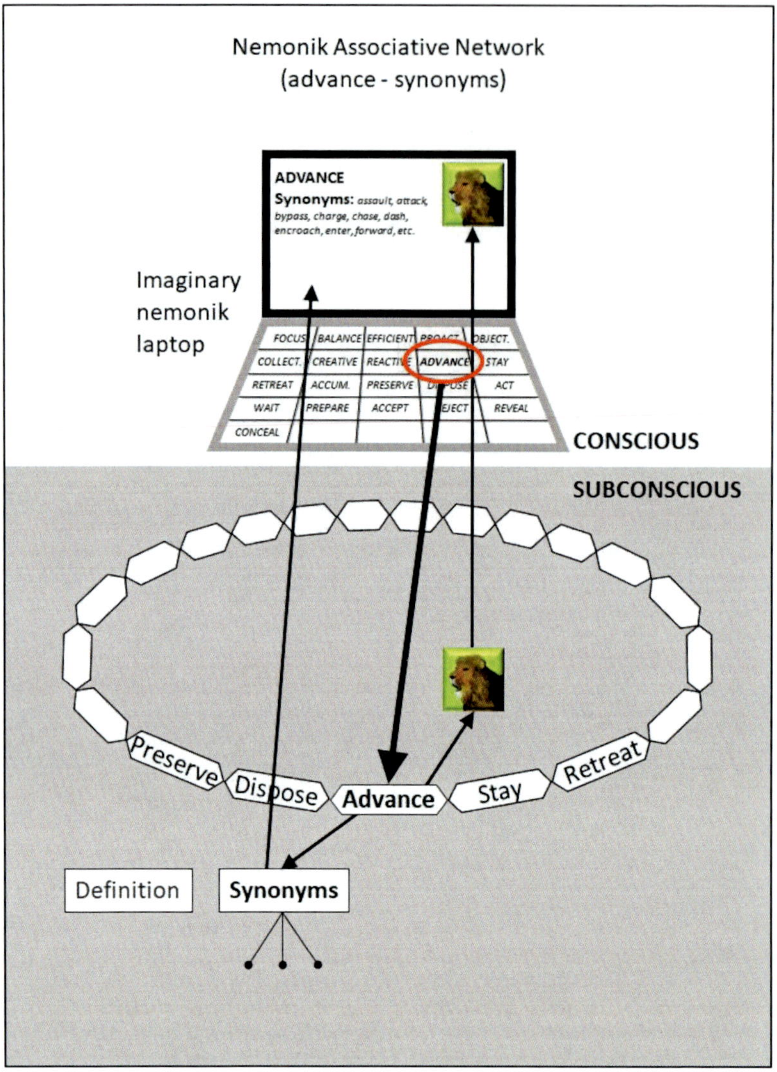

Figure 8: Advance and synonyms.

Table 3: Study nemoniks and their synonyms.

	FOCUS—concentrate, goal, aim, attitude, devote, direction, mission, purpose, resolution, single-minded, target, etc.
	BALANCE—equilibrium, harmony, tranquil, calm, modest, quiet, serene, stability, still, symmetry, synchronize, etc.
	EFFICIENT—effortless, practised, capable, competent, experience, fast, knowhow, rapid, skill, etc.
	PROACTIVE—prevent, avoid, avert, early, foil, forestall, precipitate, preclude, pre-emptive, ward-off, etc.
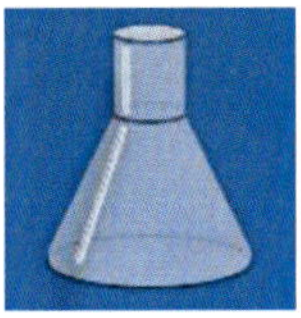	OBJECTIVE—science, laws, detached, impartial, facts, logic, rational, truth, unbiased, unemotional, unprejudiced, etc.
	COLLECTIVE—group, rules, community, bureaucracy, cooperate, enforce, indoctrinate, morals, rational, etc.

Table 3: Study nemoniks and their synonyms.

	CREATIVE—future, imagination, idea, inspiration, invention, original, progress, unique, vision, etc.
	REACTIVE—past, skill, automatic, competence, emotion, experience, feeling, habit, perfection, reflex, speed, etc.
	ADVANCE—attack, bypass, encroach, forward, infiltrate, invade, offensive, progress, vanguard, etc.
	STAY—defend, entrench, stationary, immobile, inert, motionless, rest, rooted, stand, unmoving, unyielding, etc.
	RETREAT—withdraw, fallback, rout, flee, pullback, defeat, demote, escape, evacuate, exit, resign, vacate, etc.
	ACCUMULATE—amass, collect, acquire, earn, gain, gather, incorporate, merge, procure, raid, seize, take, etc.

Table 3: Study nemoniks and their synonyms.	
	PRESERVE—conserve, repair, hold, keep, maintain, regroup, renovate, rescue, restore, save, shelter, upkeep, etc.
	DISPOSE—discard, abolish, annihilate, cull, delete, demolish, destroy, eliminate, eradicate, obliterate, purge, etc.
	ACT—do, execute, perform, strike, tackle, decisive, assertive, assured, commanding, determined, rash, etc.
	WAIT—adjourn, delay, rest, suspend, postpone, recess, patience, hesitant, indecisive, irresolute, reluctant, etc.
	PREPARE—positioning, plan, prioritize, organize, allocate, coordinate, decide, manage, delegate, exercise, etc.
	ACCEPT—approve, agree, consent, sanction, endorse, permit, trust, comply, convinced, gullible, naive, etc.

Table 3: Study nemoniks and their synonyms.	
	REJECT—disapprove, disagree, refuse, object, distrust, decline, condemn, censure, prohibit, ban, cancel, inhibit, etc.
	REVEAL—disclose, divulge, announce, explain, honest, publish, truth, tell, unveil, leak, inform, clarify, enlighten., etc.
	CONCEAL—lie, dishonest, hide, blind sight, feign, cover-up, deceive, disguise, ambush, entrap, camouflage, etc.

Test Nemoniks and Synonyms

Instructions

Write the nemonik for each set of synonyms and its corresponding number on a piece of paper. The correct answers will be on the page after the questions.

Name the nemonik for each definition.	
1	Amass, collect, acquire, earn, gain, gather, incorporate, merge, procure, raid, seize, take.
2	Approve, agree, consent, sanction, endorse, permit, trust, comply, convinced, gullible, naive.
3	Attack, bypass, encroach, forward, infiltrate, invade, offensive, progress, vanguard.
4	Concentrate, goal, aim, attitude, devote, direction, mission, purpose, resolution, single-minded, target.
5	Conserve, repair, hold, keep, maintain, regroup, renovate, rescue, restore, save, shelter, upkeep.
6	Defend, entrench, stationary, immobile, inert, motionless, rest, rooted, stand, unmoving, unyielding.
7	Adjourn, delay, rest, suspend, postpone, recess, patience, hesitant, indecisive, irresolute, reluctant.
8	Disapprove, disagree, refuse, object, distrust, decline, condemn, censure, prohibit, ban, cancel, inhibit.

Answers: nemonik for each set of synonyms.	
1	Accumulate.
2	Accept.
3	Advance.
4	Focus.
5	Preserve.
6	Stay.
7	Wait.
8	Reject.

Name the nemonik for each definition.	
9	Discard, abolish, annihilate, cull, delete, demolish, destroy, eliminate, eradicate, obliterate, purge.
10	Disclose, divulge, announce, explain, honest, publish, truth, tell, unveil, leak, inform, clarify, enlighten.
11	Do, execute, perform, strike, tackle, decisive, assertive, assured, commanding, determined, rash.
12	Effortless, practised, capable, competent, experience, fast, knowhow, rapid, skill.
13	Equilibrium, harmony, tranquil, calm, modest, quiet, serene, stability, still, symmetry, synchronize.
14	Future, imagination, idea, inspiration, invention, original, progress, unique, vision.
15	Group, rules, community, bureaucracy, cooperate, enforce, indoctrinate, morals, rational.
16	Lie, dishonest, hide, blind sight, feign, cover-up, deceive, disguise, ambush, entrap, camouflage.
17	Past, skill, automatic, competence, emotion, experience, feeling, habit, perfection, reflex, speed.
18	Positioning, plan, prioritize, organize, allocate, coordinate, decide, manage, delegate, exercise.
19	Prevent, avoid, avert, early, foil, forestall, precipitate, preclude, pre-emptive, ward-off.

Answers: nemonik for each set of synonyms..	
9	Dispose.
10	Reveal.
11	Act.
12	Efficient
13	Balance.
14	Creative.
15	Collective.
16	Conceal.
17	Reactive.
18	Prepare.
19	Proactive.

Name the nemonik for each definition.	
20	Science, laws, detached, impartial, facts, logic, rational, truth, unbiased, unemotional, unprejudiced.
21	Withdraw, fallback, rout, flee, pullback, defeat, demote, escape, evacuate, exit, resign, vacate.

Answers: nemonik for each set of synonyms..	
20	Objective.
21	Retreat.

Before moving on to the next exercise, make sure that you are familiar with the synonyms for each nemonik. The following exercises aim to expand your Network with SWOT analyses, questions, and proverbs.

2.5 Nemoniks and SWOT analyses

The aim of this exercise is extending your subconscious *Nemonik Associative Network* (Network) by studying SWOT-analyses associated with each nemonik as described in the following Table 4.

SWOT-analyses were introduced by the American business consultant Albert S. Humphrey (1926-2005). A SWOT-analysis is a problem-solving tool that evaluates the Strengths, Weaknesses, Opportunities, and Threats of each nemonik in comparison to the actual situation. Strength is an intrinsic advantage; Weakness is an intrinsic disadvantage; Opportunity is a potential advantage; and Threat is a potential disadvantage.

A SWOT-analysis of each nemonik fosters the generation of questions and ideas associated with that nemonik. As illustrated in the following Figure 9, consciously recalling the access code [ADVANCE SWOT] will return: *"Strength: decreasing the distance to your goal. Weakness: increasing the length of your supply lines. Opportunity: obtaining resources and destroying threats. Threat: entering unknown territory and running into traps."*. Hence, the access code for your Network will lead the conscious straight to the subconscious SWOT analyses and the other problem-solving tools associated with that nemonik. It is emphasized that people and situations are different. Therefore, the presented SWOT-analyses are not complete.

Remember, you have to study those SWOT analyses until your subconscious has habituated mindsets for them. That is when your genius starts the pure magic.

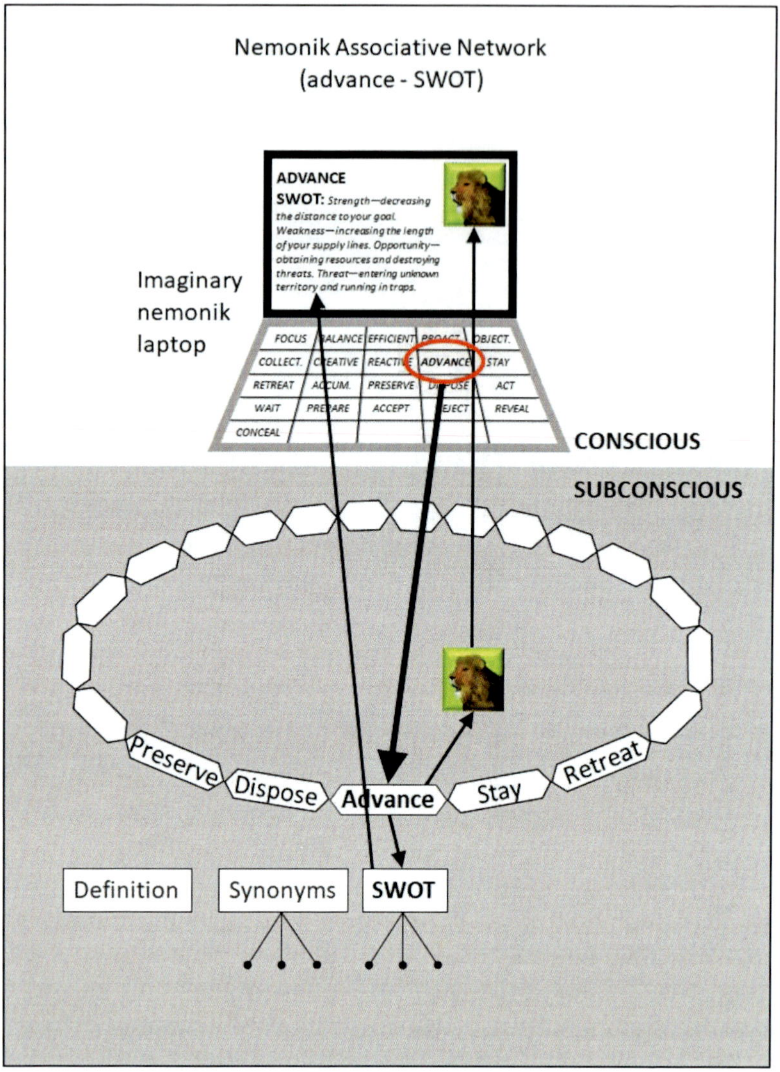

Figure 9: Advance and SWOT.

Table 4: Study nemoniks and their SWOT analyses.	
	FOCUS—Strength: concentrating all available resources on your goal. Weakness: leaving no resources for productive alternatives. Opportunity: maximizing success despite distractions. Threat: fostering static thinking in hyperdynamic reality.
	BALANCE—Strength: maintaining productive mental and physical harmony. Weakness: maintaining counterproductive mental and physical harmony. Opportunity: preventing counterproductive reactions. Threat: fostering counterproductive inactivity.
	EFFICIENT—Strength: reducing time, effort, and resources required to reach your goal. Weakness: inhibiting the focus on your goal if you go with the flow. Opportunity: preserving time, effort, and resources for future operations. Threat: alienation from other people.
	PROACTIVE—Strength: preventing problems growing out of control. Weakness: focusing on potential problems reduces the focus on your current goal. Opportunity: preserving time, effort, and resources for future problems. Threat: rash actions could destroy future opportunities.

Table 4: Study nemoniks and their SWOT analyses.	
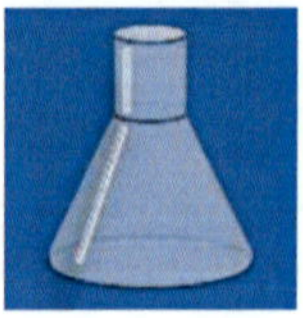	OBJECTIVE—Strength: revealing the order described by the immutable laws of nature. Weakness: objective truth is temporary, rather than eternal. Opportunity: fostering survival through scientific and technological development. Threat: inhibiting survival through scientific and technological development.
	COLLECTIVE—Strength: changeable artificial rules create order in and between groups. Weakness: enforcing the law inhibits freedom fostering rebellion. Opportunity: fostering productive cooperation. Threat: strict rules create a static collective in a dynamic reality.
	CREATIVE—Strength: fostering unique ideas dealing with the future and the unknown. Weakness: ideas and inventions are not subjected to logic. Opportunity: fostering progress and deal with non-factual problems. Threat: fostering the destruction of the planet and humanity.
	REACTIVE—Strength: providing the fastest and most accurate response possible. Weakness: producing counterproductive affecters such as anger. Opportunity: fostering productive habits and perfection. Threat: fostering counterproductive habits and repeated failures.

Table 4: Study nemoniks and their SWOT analyses.	
	ADVANCE—Strength: decreasing the distance to your goal. Weakness: increasing the length of your supply lines. Opportunity: obtaining resources and destroying threats. Threat: entering unknown territory and running into traps.
	STAY—Strength: preserving previous advances. Weakness: inhibiting the initiative, momentum, and surprise. Opportunity: fostering recuperation and regrouping. Threat: fostering predictability, isolation, immobility, and depleted supplies.
	RETREAT—Strength: fostering the preservation of life and resources. Weakness: increasing the distance to your goal. Opportunity: returning to known territory, safety, and resources. Threat: turning into a counterproductive rout.
	ACCUMULATE—Strength: increasing the material safety buffer. Weakness: increasing the preservation cost. Opportunity: fostering strategic options, safety, comfort, and freedom. Threat: fostering accumulation for the sake of accumulation.
	PRESERVE—Strength: protecting the accumulated resources. Weakness: cost will reduce accumulated resources. Opportunity: maintaining strategic options. Threat: maintaining what should be disposed.

Table 4: Study nemoniks and their SWOT analyses.	
	DISPOSE—Strength: decreasing preservation costs of accumulated resources. Weakness: losing accumulated resources. Opportunity: increased mobility. Threat: needing the disposed resources in the future.
	ACT—Strength: execution will turn your dreams into reality. Weakness: movement will evoke an opposing reaction of unpredictable magnitude. Opportunity: fostering initiative, momentum, and surprise. Threat: rashness to reduce the stress and anxiety of waiting.
	WAIT—Strength: acting at the most productive time. Weakness: inhibiting initiative, advance, and momentum. Opportunity: situation could improve without action. Threat: situation could deteriorate without action.
	PREPARE—Strength: improving future position. Weakness: decreasing current resources. Opportunity: make the impossible possible. Threat: counterproductive procrastination.
	ACCEPT—Strength: fostering the acquisition of true information. Weakness: fostering the acquisition of false information. Opportunity: fostering the trust of others. Threat: fostering gullibility and superstition.

Table 4: Study nemoniks and their SWOT analyses.	
	REJECT—Strength: inhibiting the acquisition of false information. Weakness: inhibiting the acquisition of true information. Opportunity: preventing traps and ambushes of opponents. Threat: inhibiting support of potential allies.
	REVEAL—Strength: fostering simplicity, because there is nothing to conceal. Weakness: fostering predictability. Opportunity: gaining trust. Threat: trapped and ambushed by opponents.
	CONCEAL—Strength: inhibiting predictability. Weakness: fostering complexity, which is hard to remember. Opportunity: trapping and ambushing opponents. Threat: inhibiting support of potential allies.

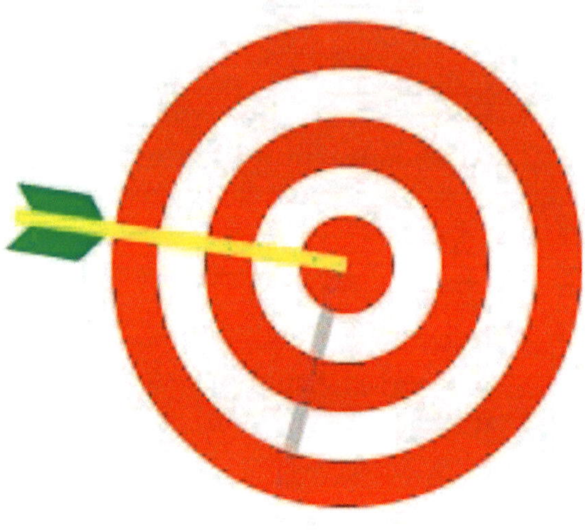

Test Nemoniks and SWOT analyses

Instructions

Write the nemonik for each SWOT analysis and its corresponding number on a piece of paper. The correct answers will be on the page after the questions.

Name the nemonik for each SWOT analysis.	
1	Strength: fostering the acquisition of true information. Weakness: fostering the acquisition of false information. Opportunity: fostering the trust of others. Threat: fostering gullibility and superstition.
2	Strength: increasing the material safety buffer. Weakness: increasing the preservation cost. Opportunity: fostering strategic options, safety, comfort, and freedom. Threat: fostering accumulation for the sake of accumulation.
3	Strength: execution will turn your dreams into reality. Weakness: movement will evoke an opposing reaction of unpredictable magnitude. Opportunity: fostering initiative, momentum, and surprise. Threat: rashness to reduce the stress and anxiety of waiting.
4	Strength: decreasing the distance to your goal. Weakness: increasing the length of your supply lines. Opportunity: obtaining resources and destroying threats. Threat: entering unknown territory and running in traps.

Answers: nemonik for each SWOT analysis.
1 Accept.
2 Accumulate.
3 Act.
4 Advance.

Name the nemonik for each SWOT analysis.

5 Strength: maintaining productive mental and physical harmony. Weakness: maintaining counterproductive mental and physical harmony. Opportunity: preventing counterproductive reactions. Threat: fostering counterproductive inactivity.

6 Strength: changeable artificial rules create order in and between groups. Weakness: enforcing the law inhibits freedom fostering rebellion. Opportunity: fostering productive cooperation. Threat: strict rules create a static collective in a dynamic reality.

7 Strength: inhibiting predictability. Weakness: fostering complexity, which is hard to remember. Opportunity: trapping and ambushing opponents. Threat: inhibiting support of potential allies.

8 Strength: fostering unique ideas dealing with the future and the unknown. Weakness: ideas and inventions are not subjected to logic. Opportunity: fostering progress and deal with non-factual problems. Threat: fostering the destruction of the planet and humanity.

9 Strength: decreasing preservation costs of accumulated resources. Weakness: losing accumulated resources. Opportunity: increased mobility. Threat: needing the disposed resources in the future.

10 Strength: reducing time, effort, and resources required to reach your goal. Weakness: inhibiting the focus on your goal if you go with the flow. Opportunity: preserving time, effort, and resources for future operations. Threat: alienation from other people.

Answers: nemonik for each SWOT analysis.	
5	Balance.
6	Collective.
7	Conceal.
8	Creative.
9	Dispose.
10	Efficient.

Name the nemonik for each SWOT analysis.	
11	Strength: concentrating all available resources on your goal. Weakness: leaving no resources for productive alternatives. Opportunity: maximizing success despite distractions. Threat: fostering static thinking in hyperdynamic reality.
12	Strength: revealing the order described by the immutable laws of nature. Weakness: objective truth is temporary, rather than eternal. Opportunity: fostering survival through scientific and technological development. Threat: inhibiting survival through scientific and technological development.
13	Strength: improving future position. Weakness: decreasing current resources. Opportunity: make the impossible possible. Threat: counterproductive procrastination.
14	Strength: protecting the accumulated resources. Weakness: cost will reduce accumulated resources. Opportunity: maintaining strategic options. Threat: maintaining what should be disposed.
15	Strength: preventing problems growing out of control. Weakness: focusing on potential problems reduces the focus on your current goal. Opportunity: preserving time, effort, and resources for future problems. Threat: rash actions could destroy future opportunities.

Answers: nemonik for each SWOT analysis.	
11	Focus.
12	Objective.
13	Prepare.
14	Preserve.
15	Proactivity.

Name the nemonik for each SWOT analysis.	
16	Strength: providing the fastest and most accurate response possible. Weakness: producing counterproductive affecters such as anger. Opportunity: fostering productive habits and perfection. Threat: fostering counterproductive habits and repeated failures.
17	Strength: inhibiting the acquisition of false information. Weakness: inhibiting the acquisition of true information. Opportunity: preventing traps and ambushes of opponents. Threat: inhibiting support of potential allies.
18	Strength: fostering the preservation of life and resources. Weakness: increasing the distance to your goal. Opportunity: returning to known territory, safety, and resources. Threat: turning into a counterproductive rout.
19	Strength: fostering simplicity, because there is nothing to conceal. Weakness: fostering predictability. Opportunity: gaining trust. Threat: trapped and ambushed by opponents.

Answers: nemonik for each SWOT analysis.	
16	Reactive.
17	Reject.
18	Retreat.
19	Reveal.

Name the nemonik for each SWOT analysis.	
20	Strength: preserving previous advances. Weakness: inhibiting the initiative, momentum, and surprise. Opportunity: fostering recuperation and regrouping. Threat: fostering predictability, isolation, immobility, and depleted supplies.
21	Strength: acting at the most productive time. Weakness: inhibiting initiative, advance, and momentum. Opportunity: situation could improve without action. Threat: situation could deteriorate without action.

Answers: nemonik for each SWOT analysis.	
20	Stay.
21	Wait.

Before moving on to the next exercise, make sure that you are familiar with the SWOT analysis for each nemonik. The following exercises aim to expand your Network with questions, and proverbs.

2.6 Nemoniks and Questions

The aim of this exercise is extending your subconscious *Nemonik Associative Network* (Network) by studying questions associated with each nemonik as described in the following Table 5.

Children use repeatedly key questions such as: Why? Where? What? Who? When? How? We should learn from the children and use their questions to evaluate each nemonik. For example, why would you advance? Where would your advance to? What resources do you intend to advance? Who is involved in your advance? When are you going to advance? How are you going to advance? Hence, those six words provide 126 potential critical questions activating your Network.[3]

Table 5 provide specific questions for each nemonik. As illustrated in the following Figure 10, consciously recalling the access code [ADVANCE QUESTIONS] will return: *"Would advancing maximize your success? Do you have the resources to advance? Are you protected during your advance? Can you sustain longer supply-lines? Can you stay in the advanced position? Do you have reserves? Would advancing change your future mobility?"* Hence, the access code will lead the conscious straight to the subconscious questions and other problem-solving tools associated with that nemonik. It is emphasized that people and situations are different. Therefore, the presented questions are not complete.

---

[3] 21 nemoniks x 6 questions = 126 potential questions.

Remember, you have to study those questions until your subconscious has habituated mindsets for them. That is when your genius starts the pure magic.

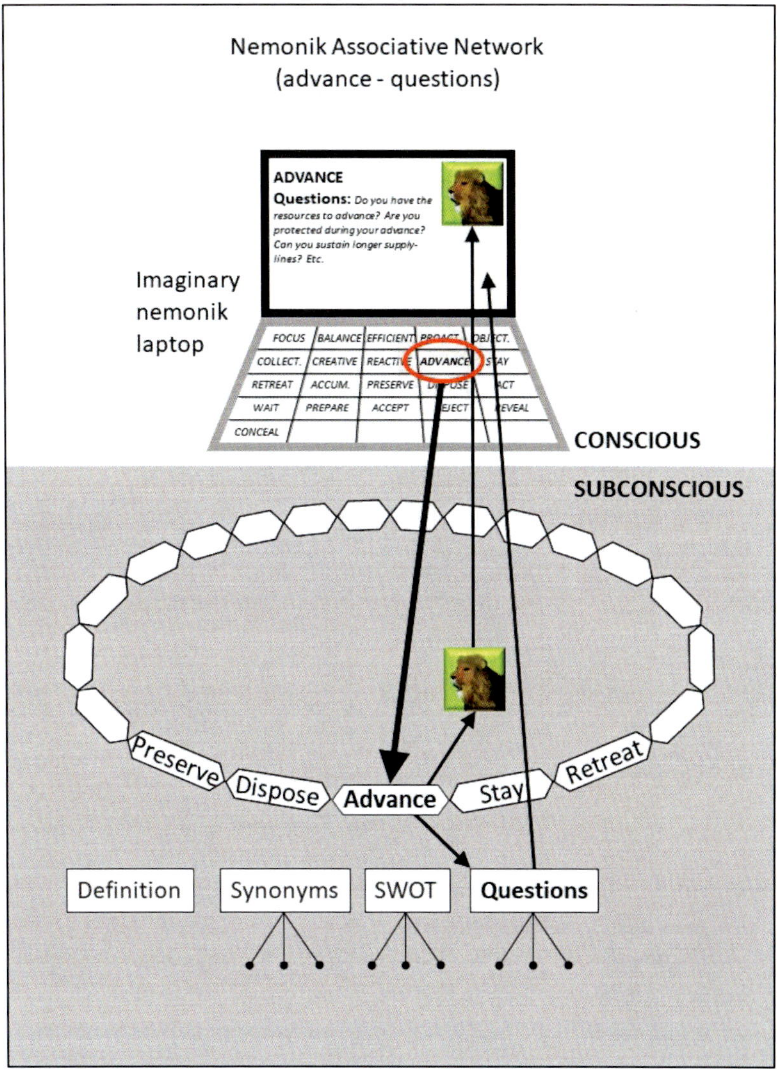

Figure 10: Advance and questions.

Table 5: Study nemoniks and their questions.

	FOCUS—Are you maximizing your success? What do you want to obtain? What do you want to escape? What is your mission statement? Have you set your priorities? etc.
	BALANCE—Do you maintain the external equilibrium? Do you maintain the internal harmony? Are you preventing chain-reactions? etc.
	EFFICIENT—Do you achieve without effort? Do you align with the forces around you? Do you gain more with the same resources? Do you obtain the same with fewer resources? etc.
	PROACTIVE—Is your action pre-emptive? Do you prevent, rather than cure? Do you act before the events take over? Do you address problems, rather than consequences? etc.
	OBJECTIVE—Do you have valid and reliable facts? Would you use logic? Will you apply the laws of nature? Do you rely on science? etc.
	COLLECTIVE—Do you have legal evidence? Would you use rhetoric? Will you apply the rules of people? Do you rely on dogmas? etc.

Table 5: Study nemoniks and their questions.	
	CREATIVE—Would you search for new ideas? Would you accept an idea if it is not yours? Will you give your genius enough time to generate ideas? Do you recall each nemonik to generate ideas? etc.
	REACTIVE—Do you practice enough to create habitual mindsets? Are you managing your genius? Do you trust your genius unconditionally? Are you able to act with a silent conscious? etc.
	ADVANCE—Do you respond to your opponent's retreat? Can you stay in your forward position? Do you have the resources to move forward? Can you protect longer supply-lines? etc.
	STAY—Is this a defendable position? Do you have the resources to defend this position? Could you stay as long as it takes? Are you able to move out when needed? Do you have protected supply-lines? etc.
	RETREAT—Do you respond to your opponent's advance? Can you stay in your backward position? Do you have the resources to move backwards? Are you shortening your supply lines? etc.
	ACCUMULATE—Do you obtain what you need, rather than what you want? Can you afford the acquisition cost? Can you preserve

Table 5: Study nemoniks and their questions.	
	what you obtain? Are the obtained resources improving your mobility? etc.
	PRESERVE—Do you maintain what you need, rather than what you want? Can you afford the maintenance cost? Is keeping more effective than disposing? Are the maintained resources improving your mobility? etc.
	DISPOSE—Are you cutting your losses? Can you afford the cost of reducing your resources? Did you preserve what you should have thrown away? Would reducing the resources improve your mobility? etc.
	ACT—Are you determined? Will your position deteriorate over time? Do you have to act? Is this the right time to act? Are you prepared for the inevitable reaction? Are you procrastinating? etc.
	WAIT—Are you patient? Is this a fight you cannot win? Will your position improve over time? Are you too early? Are you too late? Can you manage your pre-action anxiety? Are you prepared for action? Do you avoid procrastination? etc.

Table 5: Study nemoniks and their questions.	
	PREPARE—Are you positioning yourself? Do you have time left before you are forced to act? Did you set yourself time limits? Are you avoiding perfection? Do you avoid procrastination? etc.
	ACCEPT—Are you too suspicious? Do you accept what cannot be changed? Could the provided information be an opportunity? Is the provided information supported by logic? etc.
	REJECT—Are you too gullible? Is your heart detecting a lie? Could the provided information be a threat? Is the provided information only supported by rhetoric? etc.
	REVEAL—Are you cooperating? Are you using honey, rather than vinegar? Is informing others strengthening your position? Will the receivers keep the secret? etc.
	CONCEAL—Are you setting a trap? Could your threat prepare your opponent? Is informing others weakening your position? Will the receivers spread the secret? etc.

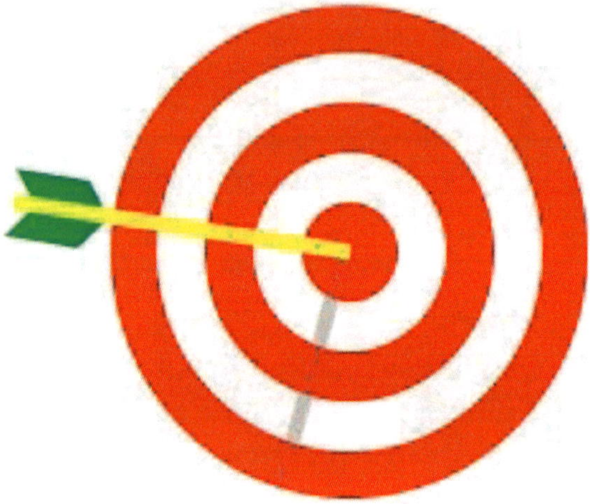

Test Nemoniks and Questions

Instructions

Write the nemonik for each set of questions and its corresponding number on a piece of paper. The correct answers will be on the page after the questions.

Name the nemonik for each set of questions.	
1	Are you cooperating? Are you using honey, rather than vinegar? Is informing others strengthening your position? Will the receivers keep the secret? etc.
2	Are you cutting your losses? Can you afford the cost of reducing your resources? Did you preserve what you should have thrown away? Would reducing the resources improve your mobility? etc.
3	Are you maximizing your success? What do you want to obtain? What do you want to escape? What is your mission statement? Have you set your priorities? etc.
4	Are you positioning yourself? Do you have time left before you are forced to act? Did you set yourself time limits? Are you avoiding perfection? Do you avoid procrastination? etc.
5	Are you setting a trap? Could your threat prepare your opponent? Is informing others weakening your position? Will the receivers spread the secret? etc.

Answers: nemonik for each set of questions.
1 Reveal.
2 Dispose.
3 Focus.
4 Prepare.
5 Conceal.

Name the nemonik for each set of questions.	
6	Are you too gullible? Is your heart detecting a lie? Could the provided information be a threat? Is the provided information only supported by rhetoric? etc.
7	Are you too suspicious? Do you accept what cannot be changed? Could the provided information be an opportunity? Is the provided information supported by logic? etc.
8	Do you achieve without effort? Do you align with the forces around you? Do you gain more with the same resources? Do you obtain the same with fewer resources? etc.
9	Do you have legal evidence? Would you use rhetoric? Will you apply the rules of people? Do you rely on dogmas? etc.
10	Do you have valid and reliable facts? Would you use logic? Will you apply the laws of nature? Do you rely on science? etc.
11	Do you maintain the external equilibrium? Do you maintain the internal harmony? Are you preventing chain-reactions? etc.
12	Do you maintain what you need, rather than what you want? Can you afford the maintenance cost? Is keeping more effective than disposing? Are the maintained resources improving your mobility? etc.

Answers: nemonik for each set of questions.	
6	Reject.
7	Accept.
8	Efficient
9	Collective.
10	Objective.
11	Balance.
12	Preserve.

Name the nemonik for each set of questions.	
13	Do you obtain what you need, rather than what you want? Can you afford the acquisition cost? Can you preserve what you obtain? Are the obtained resources improving your mobility? etc.
14	Do you practice enough to create habitual mindsets? Are you managing your genius? Do you trust your genius unconditionally? Are you able to act with a silent conscious? etc.
15	Do you respond to your opponent's advance? Can you stay in your backward position? Do you have the resources to move backwards? Are you shortening your supply lines? etc.
16	Do you respond to your opponent's retreat? Can you stay in your forward position? Do you have the resources to move forward? Can you protect longer supply-lines? etc..
17	Is this a defendable position? Do you have the resources to defend this position? Could you stay as long as it takes? Are you able to move out when needed? Do you have protected supply-lines? etc.
18	Is your action pre-emptive? Do you prevent, rather than cure? Do you act before the events take over? Do you address problems, rather than consequences? etc.
19	Are you determined? Will your position deteriorate over time? Do you have to act? Is this the right time to act? Are you prepared for the inevitable reaction? Are you procrastinating? etc.

Answers: nemonik for each set of questions.	
13	Accumulate.
14	Reactive.
15	Retreat.
16	Advance.
17	Stay.
18	Proactive.
19	Act.

Name the nemonik for each set of questions.	
20	Are you patient? Is this a fight you cannot win? Will your position improve over time? Are you too early? Are you too late? Can you manage your pre-action anxiety? Are you prepared for action? Do you avoid procrastination? etc.
21	Would you search for new ideas? Would you accept an idea if it is not yours? Will you give your genius enough time to generate ideas? Do you recall each nemonik to generate ideas? etc.

Answers: nemonik for each set of questions.	
20	Wait.
21	Creative.

Before moving to the next exercise, make sure that you are familiar with the potential questions for each nemonik. The following exercises aim to expand your Network with proverbs.

2.7 Nemoniks and Proverbs

The aim of this exercise is extending your subconscious *Nemonik Associative Network* (Network) by studying proverbs associated with each nemonik as described in the following Table 6. Knowledge is a set of valid and reliable descriptions of reality, while wisdom is the productive application of that knowledge.

Any fool could learn to split an atom,
but only a wise person knows what to do with it.

The wisdom of humanity is accumulated in proverbs. Your ancestors paid with pain, blood, and tears for that information. Therefore, only fools ignore that treasure throve.

As illustrated in the following Figure 11, for nemonik thinkers, consciously recalling the access code [ADVANCE PROVERBS] will return: *"There is no true advance without a goal. Advance only to places where you can stay. An overextended advance is a failed advance. Apply your greatest strength to the greatest weakness. Advancing downhill adds gravity to momentum. etc."* Hence, the access code will lead the conscious straight to the subconscious proverbs and other problem-solving tools associated with that nemonik. It is emphasized that the presented proverbs are not complete.

Remember, you have to study those proverbs until your subconscious has habituated mindsets for them. That is when your genius starts the pure magic.

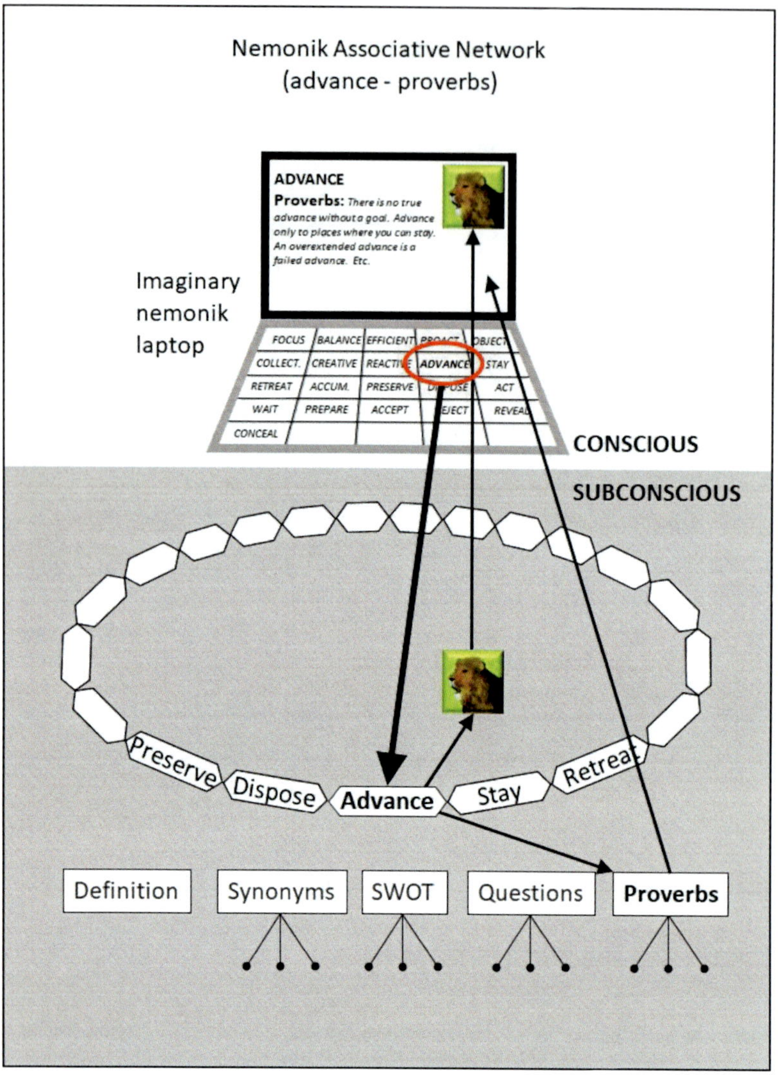

Figure 11: Advance and proverbs.

Table 6: Study nemoniks and their proverbs.	
	FOCUS—Obtain what you seek and escape what you suffer. Before hitting the target, you must select one. Ask for nothing and you will get nothing. Aim higher than the target, because the arrow drops. Concentrate on success, rather than winning. Don't let others determine your goals. Concentrate on things that work for you. Concentrate on goals, rather than obstacles. Aim on what is, rather than what should be. Don't sweat the small stuff. etc.
	BALANCE—Don't rock the boat. Moderation is the secret to lasting success. No bird flies on one wing. No rose is without thorns. Nature equalizes without favouritism. Moderating forces destroy extremes. Too much is as bad as too little. What goes up must come down. What goes around comes around. Live not only for yourself or only for others. etc.
	EFFICIENT—Maximize success with minimum resources. Act with competence and work without effort. Follow the way of least resistance. Work smarter, rather than harder. Stop doing what is not working. If it is not broken, don't fix it. Don't grasp a sword by the blade. Don't cut the branch you sit on. Don't try to reinvent the wheel. Let sleeping dogs lie. etc.

Table 6: Study nemoniks and their proverbs.	
	PROACTIVE—Sages do now, what fools do later. A neglected duty returns with seven others. Prevention is the best cure. Better to prevent than to regret. Act when it has not happened yet. Turn threats early into opportunities. The sooner begun, the sooner done. The early bird has the worm. Lock the stable-door before the horse is gone. Threats might be opportunities in disguise. etc.
	OBJECTIVE—Self-doubt is the beginning of true knowledge. Neutrality is the child of reason and logic. Reason and logic succeed where force fails. Convince with reason, rather than dogmas. Focus on facts, rather than rhetoric. Check the fact before you act. Eliminate all alternatives and the truth remains. Absence of evidence is no evidence of absence. Garbage in garbage out. etc.
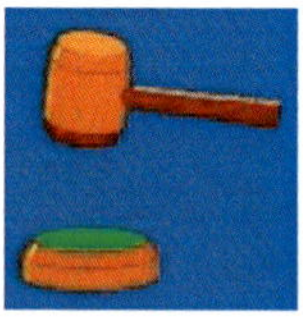	COLLECTIVE—United we stand, divided we fall. Do to others as you like others do to you. Justice is harder to define than injustice. All are equal in law. Innocent until proven guilty. Reasonable doubt is the measure of law. Ignorance of the law is no legal defence. Right carried too far becomes wrong. Possession is nine-tenth of the law. If people ignore authority, then greater authority will rise. etc.

Table 6: Study nemoniks and their proverbs.	
	CREATIVE—Don't repeat things that fail. Only new ideas can solve old problems. If ideas are not different, then they are not creative. A good idea remains good; even if it is not yours. Be patient; creativity takes time. A silent mind fosters imagination. Ideas arrive with early daylight. Ninety per cent of inspiration is perspiration. It is a thin line between fantasy and insanity. Leave criticizing your ideas to others. etc.
	REACTIVE—Practice makes perfect. Good habits make an easy life. Act with a silent mind. Intuition might be wrong, but it never lies. Fear is the killer of the little mind. Hate is a fire that will consume you. Arrogance goes before the fall. Love is a game for two that both can win. The heart has reason not understood by reason. Night brings counsel. etc.
	ADVANCE—Progress only to places where you can stay. An overextended move forward is a failed move. Apply your greatest strength to the greatest weakness. Going forward downhill adds gravity to momentum. Attack is the best defence. Surprise is the core of a strong advance. Leave an easy retreat for your opponent. Use your opponent's retreat to move forward. Preserve your supply lines while moving forward. etc.

Table 6: Study nemoniks and their proverbs.	
	STAY—Better not to move at all, than do so and regret it. If you leave your place, then you lose it. The absent person is always wrong. Plants often moved never thrive. When the cat is away the mice will play. Only lodge there where you can move. Those who do not move will obtain. A rolling stone gathers no moss. Water that has reached its level does not flow. Though the wind blows, the mountain stays. etc.
	RETREAT—It is no disgrace to flee from a tiger. Better to flee and live than to die a hero. Withdraw to fight another day. Return towards your strength. Fight when you can win; withdraw when you cannot. Withdraw faster than your opponent advances. A withdrawal without rout is no defeat. Better to return halfway than to continue in the wrong direction. Quit while you are ahead. Absence makes the heart grow fonder. etc.
	ACCUMULATE—Know what is enough and there is always enough. Don't mistake your wants for needs. What is disposed is easy to acquire. One's loss is another one's gain. Little by little makes a mountain. Small winnings make a heavy purse. A heavy purse makes a light heart. Where there is muck there is gold. Money makes money. Those who borrow live in sorrow. etc.

Table 6: Study nemoniks and their proverbs.	
	PRESERVE—What is worth accumulating is worth maintaining. Waste not, want not. A penny saved is a penny earned. Protect the goose that lays golden eggs. Maintain your shoes until you have new ones. Many tunes are played on an old fiddle. Those who don't repair their gutters have to repair their house. It is never too late to mend. Any port in a storm. Don't cut the branch you are sitting on. etc.
	DISPOSE—Hang on tightly and let go lightly. Easy come easy go. The past is water under the bridge. It is useless to flog a dead horse. A broken mirror will not reflect again. Cut your losses, before the losses cut you. You must lose a fly to catch a trout. Let the water you cannot drink flow by. Give quickly and you give twice. Only a fool will not discard a fool. etc.
	ACT—Do today what can be done today. Strike when the iron is hot. Make hay when the sun shines. Execute if the situation is deteriorating. Unless you move, nothing happens. Tide and time wait for no one. Execution turns dreams into reality. Nothing ventured, nothing gained. Fortune favours the brave. Procrastination is a thief of time. etc.

Table 6: Study nemoniks and their proverbs.	
	WAIT—Delay if the situation is improving. Many problems are none if left alone. All things come to those who do not move. Sleep on it; no matter how urgent. Fools do sooner what sages do later. Haste makes waste. Rashness shortens the road to disaster. A moment of impatience ruins your life. Each action evokes a reaction. Don't pluck flowers if you enjoy fruits. etc.
	PREPARE—The road to success is always under construction. A plan is a dream with a time limit. Planning reveals future pitfalls. Plan for war in order to preserve peace. Plan for the future, because you will be there the rest of your life. Hope for the best, but plan for the worst. A good beginning makes a good ending. Sharpen your axe before cutting a tree. Measure twice, but cut once. The need of perfection inhibits action. etc.
	ACCEPT—Agree with the immutable laws of nature. Agree with what cannot be rejected. Appreciate the bitter truth, rather than sweet lies. Foster rational thinking and inhibit rationalizations. Foster justice and inhibit justifications. Foster critical thinking and inhibit criticism. Live with what you cannot change. Slips of the tongue reveal the truth. Many truths are spoken in jest. Children and fools tell the truth. etc.

Table 6: Study nemoniks and their proverbs.	
	REJECT—Let your heart detect the lies. Be aware of the hidden snare. Flattery is velvet hiding the dagger. Those promising everything promise nothing. All that glitters is not gold. Too good to be true is not true. Refuse all you hear and half what you see. It is not true just because everyone says so. Focus on actions, rather than rhetoric. Better safe than sorry. etc.
	REVEAL—Any publicity is good publicity. One picture is worth ten thousand words. You catch more flies with honey than vinegar. Examples are better than commands. Your questions reduce your ignorance. The truth is simple because it describes reality. A fault confessed is half redressed. A bad excuse is better than none. The words 'almost' and 'about' prevent lying. Fake it, if you can't make it. etc.
	CONCEAL—Two can keep a secret if one is dead. Confess and be hanged. Keep your cards to your chest. Not all truth should be spoken. Don’t tell tales out of school. Good wine needs no advertising. Don't wash your dirty linen in public. Never show your teeth if you cannot bite. Making threats prepares your opponent. A silent fool could pass for a sage. etc.

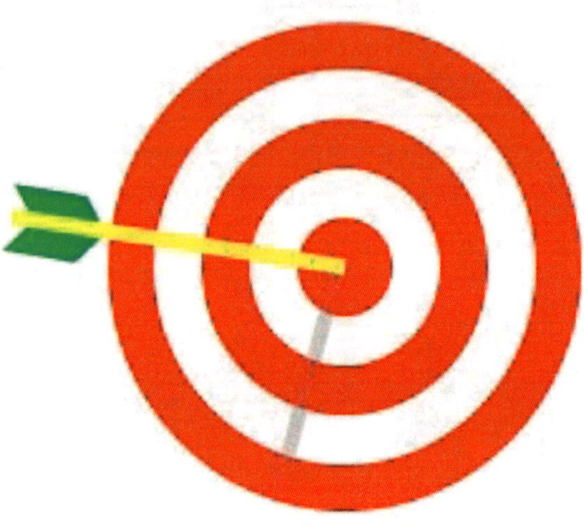

Test Nemoniks and Proverbs

Instructions

Write the nemonik for each set of proverbs and its corresponding number on a piece of paper. The correct answers will be on the page after the questions.

Name the nemonik for each set of proverbs.
1 It is no disgrace to flee from a tiger. Better to flee and live than to die a hero. Withdraw to fight another day. Return towards your strength. Fight when you can win; withdraw when you cannot. Withdraw faster than your opponent advances. A withdrawal without rout is no defeat. Better to return halfway than to continue in the wrong direction. Quit while you are ahead. Absence makes the heart grow fonder.
2 Agree with the immutable laws of nature. Agree with what cannot be rejected. Appreciate the bitter truth, rather than sweet lies. Foster rational thinking and inhibit rationalizations. Foster justice and inhibit justifications. Foster critical thinking and inhibit criticism. Live with what you cannot change. Slips of the tongue reveal the truth. Many truths are spoken in jest. Children and fools tell the truth.

Answers: nemonik for each set of proverbs.	
1	Retreat.
2	Accept.

Name the nemonik for each set of proverbs.
3 Do today what can be done today. Strike when the iron is hot. Make hay when the sun shines. Execute if the situation is deteriorating. Unless you move, nothing happens. Tide and time wait for no one. Execution turns dreams into reality. Nothing ventured, nothing gained. Fortune favours the brave. Procrastination is a thief of time.   4 Any publicity is good publicity. One picture is worth ten thousand words. You catch more flies with honey than vinegar. Examples are better than commands. Your questions reduce your ignorance. The truth is simple because it describes reality. A fault confessed is half redressed. A bad excuse is better than none. The words 'almost' and 'about' prevent lying. Fake it, if you can't make it.   5 Obtain what you seek and escape what you suffer. Before hitting the target, you must select one. Ask for nothing and you will get nothing. Aim higher than the target, because the arrow drops. Concentrate on success, rather than winning. Don't let others determine your goals. Concentrate on things that work for you. Concentrate on goals, rather than obstacles. Aim on what is, rather than what should be. Don't sweat the small stuff.

Answers: nemonik for each set of proverbs.
3 Act
4 Reveal.
5 Focus.

Name the nemonik for each set of proverbs.

6 Don't repeat things that fail. Only new ideas can solve old problems. If ideas are not different, then they are not creative. A good idea remains good; even if it is not yours. Be patient; creativity takes time. A silent mind fosters imagination. Ideas arrive with early daylight. Ninety per cent of inspiration is perspiration. It is a thin line between fantasy and insanity. Leave criticizing your ideas to others.

7 Hang on tightly and let go lightly. Easy come easy go. The past is water under the bridge. It is useless to flog a dead horse. A broken mirror will not reflect again. Cut your losses, before the losses cut you. You must lose a fly to catch a trout. Let the water you cannot drink flow by. Give quickly and you give twice. Only a fool will not discard a fool.

8 The road to success is always under construction. A plan is a dream with a time limit. Planning reveals future pitfalls. Plan for war in order to preserve peace. Plan for the future, because you will be there the rest of your life. Hope for the best, but plan for the worst. A good beginning makes a good ending. Sharpen your axe before cutting a tree. Measure twice, but cut once. The need of perfection inhibits action.

9 Know what is enough and there is always enough. Don't mistake your wants for needs. What is disposed is easy to acquire. One's loss is another one's gain. Little by little makes a mountain. Small winnings make a heavy purse. A heavy purse makes a light heart. Where there is muck there is gold. Money makes money. Those who borrow live in sorrow.

Answers: nemonik for each set of proverbs.
6 Creative.
7 Dispose.
8 Prepare.
9 Accumulate.

Name the nemonik for each set of proverbs.

10 Maximize success with minimum resources. Act with competence and work without effort. Follow the way of least resistance. Work smarter, rather than harder. Stop doing what is not working. If it is not broken, don't fix it. Don't grasp a sword by the blade. Don't cut the branch you sit on. Don't try to reinvent the wheel. Let sleeping dogs lie.

11 Don't rock the boat. Moderation is the secret to lasting success. No bird flies on one wing. No rose is without thorns. Nature equalizes without favouritism. Moderating forces destroy extremes. Too much is as bad as too little. What goes up must come down. What goes around comes around. Live not only for yourself or only for others.

12 Better not to move at all, than do so and regret it. If you leave your place, then you lose it. The absent person is always wrong. Plants often moved never thrive. When the cat is away the mice will play. Only lodge there where you can move. Those who do not move will obtain. A rolling stone gathers no moss. Water that has reached its level does not flow. Though the wind blows, the mountain stays.

13 Practice makes perfect. Good habits make an easy life. Act with a silent mind. Intuition might be wrong, but it never lies. Fear is the killer of the little mind. Hate is a fire that will consume you. Arrogance goes before the fall. Love is a game for two that both can win. The heart has reason not understood by reason. Night brings counsel.

Answers: nemonik for each set of proverbs.	
10	Efficient.
11	Balance.
12	Stay.
13	Reactive.

Name the nemonik for each set of proverbs.	
14	Let your heart detect the lies. Be aware of the hidden snare. Flattery is velvet hiding the dagger. Those promising everything promise nothing. All that glitters is not gold. Too good to be true is not true. Refuse all you hear and half what you see. It is not true just because everyone says so. Focus on actions, rather than rhetoric. Better safe than sorry.
15	Sages do now, what fools do later. A neglected duty returns with seven others. Prevention is the best cure. Better to prevent than to regret. Act when it has not happened yet. Turn threats early into opportunities. The sooner begun, the sooner done. The early bird has the worm. Lock the stable-door before the horse is gone. Threats might be opportunities in disguise.
16	Self-doubt is the beginning of true knowledge. Neutrality is the child of reason and logic. Reason and logic succeed where force fails. Convince with reason, rather than dogmas. Focus on facts, rather than rhetoric. Check the fact before you act. Eliminate all alternatives and the truth remains. Absence of evidence is no evidence of absence. Garbage in garbage out. Only reason and logic inhibit incorrect beliefs.

Answers: nemonik for each set of proverbs.	
14	Reject.
15	Proactive.
16	Objective.

Name the nemonik for each set of proverbs.

17 . Progress only to places where you can stay. An overextended move forward is a failed move. Apply your greatest strength to the greatest weakness. Going forward downhill adds gravity to momentum. Attack is the best defence. Surprise is the core of a strong advance. Leave an easy retreat for your opponent. Use your opponent's retreat to move forward. Preserve your supply lines while moving forward.

18 . Two can keep a secret if one is dead. Confess and be hanged. Keep your cards to your chest. Not all truth should be spoken. Don't tell tales out of school. Good wine needs no advertising. Don't wash your dirty linen in public. Never show your teeth if you cannot bite. Making threats prepares your opponent. A silent fool could pass for a sage.

19 . United we stand, divided we fall. Do to others as you like others do to you. Justice is harder to define than injustice. All are equal in law. Innocent until proven guilty. Reasonable doubt is the measure of law. Ignorance of the law is no legal defence. Right carried too far becomes wrong. Possession is nine-tenth of the law. If people ignore authority, then greater authority will rise.

Answers: nemonik for each set of proverbs.	
17	Advance.
18	Conceal.
19	Collective.

Name the nemonik for each set of proverbs.	
20	Delay if the situation is improving. Many problems are none if left alone. All things come to those who do not move. Sleep on it; no matter how urgent. Fools do sooner what sages do later. Haste makes waste. Rashness shortens the road to disaster. A moment of impatience ruins your life. Each action evokes a reaction. Don't pluck flowers if you enjoy fruits.
21	What is worth accumulating is worth maintaining. Waste not, want not. A penny saved is a penny earned. Protect the goose that lays golden eggs. Maintain your shoes until you have new ones. Many tunes are played on an old fiddle. Those who don't repair their gutters have to repair their house. It is never too late to mend. Any port in a storm. Don't cut the branch you are sitting on.

Answers: nemonik for each set of proverbs.	
20	Wait.
21	Preserve.

Before moving on to the next exercise, make sure that you are familiar with the proverbs associated with each nemonik.

3.0 EXERCISES MIND MANAGEMENT

Mind Management is the part of Nemonik Thinking that maximizes internal success with Concentrated-Relaxation and Mantras.

The unmanaged mind is like a wild horse that has no focus and goes wherever it wants to go on the spur of the moment. It is easily distracted and changes direction without any reason. It jumps where it should tread carefully. It follows desires, rather than needs. Unconnected thoughts come and go on the screen of the conscious creating a chaos during the day that keeps the victim awake at night. Such a mental turmoil affects the surrounding reality and its counterproductive feedback leaves internally a trail of anxiety, confusion, depression, and frustration.

Some sufferers try to escape from the chaos by taking a variety of chemicals that change their perception of reality. Others swap their unproductive coping strategies for equally unproductive philosophies. They hide in the now, at the cost of the past that has created them and from a future that could provide them with hope and opportunities. Too scared to face where they came from and too much disappointed to care where they are going. Many create as much noise as possible to mask the increasingly louder feedback signals from the protesting reality.

In one way or another, the escapees chose a suspended animation above the turmoil of real life. Nevertheless, reality pulls them inevitably back into orbit because their behaviour does not improve their internal reality in a productive way.

Underlying their turmoil is a lack of proper mind management.

If you do not take the time and effort to understand your horse, then you should not complain about a rough ride. To enjoy the ride, the horse has to be taught with patience, understanding, and kindness how to respond. Similarly, you need to teach your mind how to respond to the external reality.

Nemonik Thinking provides the tools, because it is based on a dynamic model of the mind. Furthermore, the *Associative Nemonik Network* (Network) helps to interact with the external reality in a productive way, because the twenty-one nemoniks include all the perceived aspects of Lao Zi's cosmic principle, mind, reality, and the interaction between mind and reality. It provides a simple map comprising landmarks for navigating the infinite maze of details that comprises reality. It is up to you to manage your mind with the patience, understanding, and the kindness that it deserves.

3.1 Concentrated-Relaxation

As mentioned previously, Concentrated-Relaxation is a cognitive tool that evokes semiconscious dominance by reducing conscious dominance without falling asleep. Semiconscious dominance balances conscious and subconscious dominance in order to open the communication channel between the conscious manager and the subconscious genius. A mantra is a word or phrase that is repeated in the mind during semiconscious dominance in order to modify and create mindsets. Mindsets are habituated set of rules or algorithms that generate affecters. Affecters are mental signal generated by a mindset that influences the conscious. Concentrated-Relaxation training includes: muscle and breathing awareness; evoking semiconscious dominance; progressive muscle relaxation and breathing; deepening semiconscious; creating mindsets with mantras; returning conscious dominance; and mind-sliding.

Sliding

In order to manage the mind, one has to learn to slide from one mental state to another. As illustrated in the following Figure 12, sliding is the mental process of shifting consciously the dominance window across the mental continuum. Concentration and relaxation move the dominance window respectively to the conscious and subconscious extremes of that continuum. In contrast, Concentrated-Relaxation fosters a semiconscious dominance that balances those extremes. In that way, it opens the communication channel between the conscious and subconscious.

If you want to learn new tasks, provide direction to your life, and manage the subconscious, then you have to concentrate in order to slide to conscious dominance. On the other hand, if you need mental and physical recuperation, then you should relax in order to slide to subconscious dominance. Furthermore, if you decide to create mindsets and improve your affecters, then you should slide to semiconscious dominance with Concentrated-Relaxation.

Semiconscious dominance

Semiconscious dominance is crucial for Mind Management. It is a natural mental state and everyone experiences it each day in one way or another. Hence, there is nothing mysterious about it. Semiconscious dominance could range from a light relaxed mental state to a deep trance. During semiconscious dominance, your conscious awareness of the sensory reality might change. People use a wide array of words to describe this mental state such as alpha, calm, contemplative, cool, dozy, drowsy, empty mind, enlightened, high, inspirational, introspective, meditative, peaceful, relaxed, satori, serene, silent mind, spaced-out, trance, tranquil, twilight, zone, etc.

Semiconscious dominance occurs involuntarily during absent-mindedness, falling asleep, reading, relaxation, watching a video, repetitive tasks, snoozing, waking up, etc. It can also occur involuntarily when you are wide-awake and your subconscious reactive mindmode takes control in order to cope with an emergency. For example, your subconscious might have hit the brakes of your car before you are even consciously aware of any danger. In such emergencies, your subconscious acts and your conscious becomes an observer. This is the action with a silent mind as strived for in Eastern martial arts. Hence, semiconscious dominance might be either a state of tranquillity or high-octane action.

Dominance window

Mind	
CONSCIOUS	**SUBCONSCIOUS**
Concentrate Conscious dominance Learn new tasks	
Concentrated-Relaxation Semiconscious dominance Create mindsets	
	Relax Subconscious dominance Recuperate

Figure 12: Sliding dominance window.

The preceding Figure 12, visualizes the sliding window of mental dominance. That hypothetical window determines the mental state by sliding over the mental continuum, which ranges from the conscious to the subconscious. The figure shows that semiconscious dominance is crucial for maximizing your success, because it is the communication channel between your conscious and subconscious. Semiconscious dominance allows you to access consciously subconscious information, maximize mental speed and accuracy, deploy subconscious mindpower, modify counterproductive mindsets, and foster intentionally the reception of creative and reactive affecters.

Concentration and relaxation slide the dominance window in opposite directions along the mental continuum. The more you experience your subconscious, the less you experience your conscious and vice versa. When you wake up and concentrate, then your dominance window slides from the subconscious to the conscious. In contrast, when you relax and fall asleep the opposite occurs. Hence, both concentration and relaxation could evoke periods of semiconscious dominance during the slide from one mental extreme to the other.

The conscious is mainly the mono-tasking part of your mind. In contrast, the larger subconscious is predominantly the multi-tasking part. The subconscious can activate several mindsets without conscious involvement. For instance, your subconscious can simultaneously operate a car, scratch your face, follow the GPS-directions, brake for a pedestrian crossing the road, and talk to a passenger.

The usefulness of the conscious and subconscious depends entirely on the situation. If there is no adequate mindset for the task, then you need conscious dominance and concentration in order to habituate one. If there is an adequate mindset, then semiconscious dominance might be more efficient. In that case, you need a silent conscious, because its critical approach will get in the way of activating efficiently the particular mindset.

Affecters

Conventional thinkers try to change their counterproductive affecters with conscious willpower. For example, they use it to remove bad habits, curb their desires, or evoke productive feelings. However, willpower requires concentration, which fosters conscious dominance. As illustrated in the preceding Figure 12, that dominance will inhibit the improvement of mindsets. Instead, willpower fosters a counterproductive conflict between the conscious and subconscious. Consequently, no amount of willpower could improve the affecters. Eventually, conventional thinkers just run out of willpower. That failure causes more stress, depression, frustration, and anger.

In contrast, nemonik thinkers focus on their mindsets, rather than their affecters. They use mind management, rather than mind control. They use Concentrated-Relaxation, rather than willpower. Concentrated-Relaxation uses unvoiced mantras to create productive mindsets during semiconscious dominance. Those mindsets will produce automatically productive affecters. Concentrated-Relaxation can be hidden from other people. Therefore, it can be used everywhere and under all conditions.

3.2 Mantras

As you know, a mantra is a word or phrase that is consciously repeated during semiconscious dominance in order to manage the subconscious genius. I do not know you or your circumstances. Therefore, I have no idea what kind of mantra(s) you need. The potential mantras in the following Tables 7 and 8, might give you inspiration.

Mantras for Nemoniks	
>I focus<	<on success>
>I maintain<	<the balance>
>I go<	<with the flow>
>I<	<prevent>
>I align<	<with nature>
>I unite<	<with people>
>I search<	<for ideas>
>I trust<	<my genius>
>I advance<	<to where I can stay>
>I stay<	<where I can leave>
>I retreat<	<to where I can stay>
>I accumulate<	<productive resources>
>I preserve<	<accumulated resources>
>I dispose<	<unproductive resources>
>I act<	<with determination>
>I wait<	<for the right time>
>I prepare<	<before I wait>
>I check<	<before I accept>
>I check<	<before I reject>
>I reveal<	<to allies>
>I conceal<	<from opponents>

Table 7: Mantras for nemoniks.

Creating Mantras

Alternatively, you might want to create your own mantras. The lyrics of songs and prayers might provide inspiration. You have to determine what success means for you. Are you sure that you want to change? Are you sure that you know what to change? Take your time.

Your hidden genius is like Aladdin's genie in the bottle. You get exactly what you ask for. Hence, be very careful that your mantra reflects precisely what you want to achieve. A mantra can be a two-edged sword. If you just ask for wealth, then you do not ask to be happy. After all, you might not like what your genius will give you. You might become a billionaire living in misery. If you just ask for self-confidence, you do not ask for humbleness. You might become arrogant closing your mind and stop adjusting. If you ask for success, you do not question the means and the costs. You might have no friends left.

Be careful not to use negative words in your mantras, because they might stick in your mind and reinforce unwanted negativity in your subconscious. For example: (>I will not<|<smoke>) could be a counterproductive mantra. It might glue the word 'smoke' to your subconscious, which might foster smoking. Instead, you could use a mutually exclusive mantra: (>I want only<|<fresh air>). That would exclude smoking.

Mantras are stronger if you describe the mindset as existing, rather than as a desire. For example: (>I am<|<a champion>) is stronger than (>I want to be<|<a champion>). If you desire to be a champion, then you hope to become a champion. Hope is eternal. Hence, that might never happen. Believe in your mantras, then your genius will give you exactly what you ask for. Even if the mantra is not true yet, the result will be much better. Fake it, while you make it. Act as a champion in order to become a champion.

Mantras for Positive Affirmations	
>I am<	<a champion>
>I am<	<confident>
>I am<	<successful>
>I am<	<healthy>
>I am<	<happy>
>I am<	<calm>
>I am<	<strong>
>I am<	<great>
>I am<	<a survivor>
>I am<	<relaxed>
>I am<	<competent>

Table 8: Mantras for Positive Affirmations.

3.3 Nemonik Profile

You cannot manage something that you do not understand. Hence, you have to know how the nemoniks affect your mind before you can manage your mind properly. For that reason, shuffle the deck of nemonik cards and lay five cards on the table as illustrated in the following Figure 13. In this case—WAIT; ACT; PRESERVE; REVEAL; and PREPARE. Evaluate the five cards. Discard one card that describes you least. Put the next card from the deck next to the remaining four. Repeat the process until there are only four cards left. Write down those four remaining nemoniks, because they describe you best.

Figure 13: Lay out five nemonik cards.

Repeat the entire process. However, this time, discard each card that describes you best. Write down the four remaining nemoniks, because they describe you least. Together, those eight nemoniks comprise your nemonik profile.

SPOILER ALERT

Do not read the following debrief before you have completed your profile.

Profile debrief

Each person is different and is in a different situation. Hence, there is no perfect response to this test. You have to rely on your knowledge about yourself. I can only offer some suggestions.

If you have a nemonik bias, then you might face unnecessary threats and miss out on opportunities. Therefore, you should analyse your eight liked and disliked nemoniks very carefully. Study the icon, definition, potential questions, SWOT analyses, mantras, and proverbs for each nemonik.

Nemonik thinkers have no bias for or against any nemonik. The value of each nemonik is exclusively determined by the situation. Each nemonik is just a prompt to maximize success in a different way. All nemoniks describe a nemonik thinker equally.

If the previous exercise was relatively easy for you, then you might have a bias for or against one or more nemoniks. That is dangerous, because such a bias will foster static thinking in your dynamic reality. For example, if you are biased towards ACT, then you might find yourself sooner or later in a situation where only WAIT or PREPARE could maximize your success.

The widespread use of the equal sign (=) in mathematics, science, and our thinking supports Lao Zi's idea that the cosmic principle is balance. That balance is maintained and restored by the unstoppable generic force Qi. Therefore, any extreme will be destroyed by Qi [8] [9] [10]. Hence sages avoid extremes: *"Sages reject extremes, reject grandeur, and reject extravagance"* [1] §30. People who seek extremes are likely to fail. Therefore, nemonik thinkers avoid extremes: *"Concentrate on removing extremes"* §16. *"No greater suffering than having extreme desires. No greater misfortune than not knowing what is enough"* [1] §46.

Young people seem to be greater risk-takers than older ones. The question is whether risk-taking decreases with age, or alternatively, that the risk-takers die younger. After all: *"Three in ten are companions of life. Three in ten are companions of death. Three in ten people live extremely and move into the realm of death. What is the reason? That is because they live extremely"* [1] §50. Hence, moderation might not be a characteristic of aging, but instead, a characteristic of surviving. When getting older, the moderate people are still there to tell their tales.

Be aware that nemoniks, such as ADVANCE, ACCUMULATE, and ACT, could create counterproductive extremes. On the other hand, disliking nemoniks such as, RETREAT, DISPOSE, and WAIT, could also lead to counterproductive extremes. How many extreme likes and dislikes are there in your strategic profile? If you are biased, then consider to manage your nemoniks with Concentrated-Relaxation and Mantras.

Only proceed to the next exercise when you manage this one adequately.

3.4 Preparation

Warning: Never ever use Concentrated-Relaxation or mind-sliding in situations where you have to be alert or when relaxation could become counterproductive or could cause accidents.

As you know, Concentrated-Relaxation is a cognitive tool that evokes semiconscious dominance by reducing conscious dominance without falling asleep. Concentrated-Relaxation is powerful. Therefore, handle it with the utmost respect and care. Never use it when you need to be awake and alert. For example, do not use it when driving a car or when using machinery. You might become drowsy and cause an accident.

Worrying fosters conscious dominance, which inhibits semiconscious dominance. Therefore, make in advance a to-do-list of anything that you have to attend to after the Concentrated-Relaxation session. In that way, you do not have to worry during the session about something you might forget.

Semiconscious dominance fosters your Creative mindmode and, therefore, you might get ideas during the session. To inhibit unintentional conscious thoughts, it might be better to interrupt your session and write those ideas down. Otherwise, you might keep thinking about them. So, keep pen and paper ready.

It is easier to start your training sessions in a quiet location with dimmed lights. Switch off your phone etc. Ask others not to interrupt you or put a note on the door to that effect.

Make yourself comfortable and lay down on a bed or sit down in an easy chair. Do not try to freeze your body during any stage of the session. Just keep moving around until you feel comfortable. Otherwise, your focus will be distracted by discomfort. Immobility might decrease your body temperature. Therefore, make sure that you stay warm. If you have only limited time for your session, then it might be a good idea to set an alarm.

Concentrated-Relaxation is a mental process that fosters semi-conscious dominance by balancing both conscious and subconscious dominance. That process is intentional and, therefore, it has to be directed by the conscious. Hence, you should try to stay awake during your session. When you fall asleep, you reach subconscious dominance and the session stops automatically. Do not worry! You will wake up in the normal way. You might have needed the sleep more than the exercise. Just try again.

You can use Concentrated-Relaxation as many times and as long as you want, but do not force it. Try at least one session of ten minutes a day. Be patient—do not rush through the exercises. Repeat each exercise until you can complete it without falling asleep. It might be helpful to complete the first session of each exercise with open eyes to learn the instructions.

It is easy to relax in an ideal location with soft background music, thick carpet, comfortable furniture; humming air-conditioning; fleeting scents, and dim candlelight. However, real life is different. You need Concentrated-Relaxation most during emergencies; after an accident; wet, cold, and in the middle of a tornado; during a hostile boardroom meeting; during grief and pain; after an emotional breakup; etc.

If you rely on paraphernalia, then you might be unable to reach semiconscious dominance without them when you need it most. It does not look impressive if you can only relax in the boardroom if you sit cross-legged on the table while burning incense. Hence, pairing semiconscious dominance with paraphernalia and assuming particular body positions create serious mental weaknesses.

The following Table 9 presents a legend that prevents a verbal avalanche in the instructions for the training sessions. The symbols are just common sense.

<table>
<tr><th>Legend</th></tr>
<tr><td>>inhale slow and easy<
<exhale slow and easy>
>>inhale slow, easy, and deep<<
<<exhale slow, easy, and deep>>
| hold breath short and easy.
| | hold breath long and easy.
(...x) number of times to repeat.
[...min] repeat for approximate duration in minutes.
[A] repeat until achieved.
words silently voiced in your mind
words spoken out loud</td></tr>
</table>

Table 9: Legend Concentrated-Relaxation.

Warning: Never ever use Concentrated-Relaxation or mind-sliding in situations where you have to be alert or when relaxation could become counterproductive or could cause accidents.

3.5 Semiconscious Dominance.

Before continuing, make sure that you master the previous exercise.

Prepare as described in Exercise 3.04. As illustrated in the following Table 10, stage 1 of the current exercise aims to raise your awareness of your current muscle and breathing tension. Stage 2 aims to slide your mental state from conscious dominance to semiconscious dominance without falling asleep. If you fall asleep or lose track just start again. Don't worry. You might need your sleep more than the exercise. Take it easy—go with the flow. It should be fun! Stage 6 aims to return you from semiconscious to conscious dominance. The stages 3, 4, and 5 will be discussed later.

If you are distracted during a session, do not worry. That is no surprise, because you are improving your mind. Just bring your mind softly back to the exercise. You cannot hurry your training. It will take time. However, if you keep trying you will succeed. Be patient with yourself. The teacher will come when the student is ready.

After the completion of each session, you should feel wide awake, alert, and great. If you are still drowsy, then repeat stage 6 until you achieve the intended result. Repeat the exercise at least once a day for a week.

<table>
<tr><th colspan="3">Semiconscious Dominance.</th></tr>
<tr><td colspan="3">Stage 1: Muscle and breathing awareness</td></tr>
<tr><td colspan="3">Close eyes</td></tr>
<tr><td>>>clench fists<<</td><td>||</td><td><<unclench fists>> (3x)</td></tr>
<tr><td>>>stretch arms & legs<<</td><td>||</td><td><<relax arms & legs>> (3x)</td></tr>
<tr><td>>>...<<</td><td>||</td><td><<...>> (3x)</td></tr>
<tr><td>>...<</td><td>|</td><td><...> (3x)</td></tr>
<tr><td colspan="3">Stage 2: Evoking semiconscious dominance</td></tr>
<tr><td>>at the count of three<</td><td>|</td><td><I am relaxed></td></tr>
<tr><td>>one<</td><td>|</td><td><I am relaxed></td></tr>
<tr><td>>two<</td><td>|</td><td><I am relaxed></td></tr>
<tr><td>>three<</td><td>|</td><td><I am relaxed></td></tr>
<tr><td>>I like<</td><td>|</td><td><this training> [5 min]</td></tr>
<tr><td colspan="3">Stage 6: Returning conscious dominance</td></tr>
<tr><td>>>at the count of one<<</td><td>||</td><td><<I am wide awake>></td></tr>
<tr><td>>>three<<</td><td>||</td><td><<wide awake>></td></tr>
<tr><td>>>two<<</td><td>||</td><td><<wide awake>></td></tr>
<tr><td>>>one<<</td><td>||</td><td><<wide awake>></td></tr>
<tr><td colspan="3">I am wide awake
My eyes are open
I feel alert and great</td></tr>
</table>

Table 10: Semiconscious Dominance - Exercise 3.5.

Only proceed to the next exercise when you master this one. Be patient with yourself.

3.6 Progressive Relaxation.

Before continuing, make sure that you master the previous exercise.

Prepare as described in Exercise 3.04. The current exercise aims to increase your relaxation with a method called 'Progressive Relaxation'.

It is all about balance. A relaxed body is incompatible with a tense mind. Therefore, relaxing the body will force the mind to relax. Focus your entire mind on the body part that you are relaxing. If you are distracted, do not worry. Just bring your mind back to the exercise. Move slowly from one body part to the next.

As illustrated in the following Table 11, during the first session of this exercise, apply only line 1 of stage 3.: (>my thumb< | <is relaxed>). The aim is to relax the thumb of your dominant hand. The thumb is a small part of your body and relaxing it is an achievable goal even when you are tense. If necessary, select another small part of your body. Repeat this session until you can relax your thumb adequately.

During the next session add line 2 of stage 3 (>my hand<|<is relaxed?). And so on, until you reach (>my body<|<is relaxed>).

Progressive Relaxation	
Stage 1: Muscle and breathing awareness	
Stage 2: Evoking semiconscious dominance	
Stage 3: Progressive muscle relaxation	
>my thumb<	<is relaxed> [A]
>my hand<	<is relaxed> [A]
>my arm<	<is relaxed> [A]
>my shoulders<	<are relaxed> [A]
>my neck<	<is relaxed> [A]
>my face<	<is relaxed> [A]
>my body<	<is relaxed> [A]
Stage 6: Returning conscious dominance	

Table 11: Progressive Relaxation - Exercise 3.6.

Only proceed to the next exercise when you manage this one adequately.

3.7 Deepening Semiconscious.

Before continuing, make sure that you master the previous exercise.

Prepare as described in Exercise 3.04. As illustrated in the following Table 12, the current exercise aims to deepen the state of semiconscious dominance as far as possible without falling asleep. During that state, the mantras will be optimal effective in habituating permanent mindsets.

Just before falling asleep, many people become aware of a feeling of warmth and heaviness. In accord, this deepening exercise features the mantra: (>I feel<|<warm and heavy>). Simultaneously, imagine that feeling of warmth and heaviness spreading through your body.

Deepening Semiconscious.
Stage 1: Muscle and breathing awareness
Stage 2: Evoking semiconscious dominance
Stage 3: Progressive muscle relaxation
Stage 4: Deepening semiconscious
>I relax<	<deeper and deeper> [A] >My body<	<is relaxed> [A] >I feel<	<warm and heavy> [A]
Stage 6: Returning conscious dominance

Table 12: Deepening Semiconscious- Exercise 3.7.

Only proceed to the next exercise when you manage this one adequately.

3.8 Creating Mindsets.

Before continuing, make sure that you master the previous exercise.

Prepare as described in Exercise 3.04. As illustrated in the following Table 13, the current exercise aims to create productive mindsets with Concentrated-Relaxation and mantras. A mindset is a habituated set of rules or algorithms that generates affecters. A mantra is a tool of Mind Management that comprises a word or phrase that is repeated in the mind during semiconscious dominance in order to create mindsets that manage the subconscious genius.

Select the mantras you need from the preceding Tables 7 and 8 or create your own. For the optimal effect, use only one mantra in a session.

Creating Mindsets.
Stage 1: Muscle and breathing awareness
Stage 2: Evoking semiconscious dominance
Stage 3: Progressive muscle relaxation
Stage 4: Deepening semiconscious
Stage 5: Creating mindsets with mantras
Select a mantra from the list [A]
Stage 6: Returning conscious dominance

Table 13: Creating Mindsets - Exercise 3.8.

Only proceed to the next exercise when you manage this one adequately.

3.9 Mind-sliding.

Before continuing, make sure that you master the previous exercise.

Prepare as described in Exercise 3.04. As illustrated in the following Table 14, after sufficient training, your reactive mindmode will create a mindset for Concentrated-Relaxation. That mindset allows you to slide almost instantaneously between conscious and semiconscious dominance. Therefore, you could bring the sessions back to a minimum. If you are adequately trained, you do not even have to close your eyes.

Mind Sliding.
Stage 2: Evoking semiconscious dominance
Stage 6: Returning conscious dominance

Table 14: Mind Sliding - Exercise 3.9.

Only proceed to the next exercise when you manage this one adequately.

3.10 Generalization

Before continuing, make sure that you master the previous exercise.

There is one task left. You have to generalize the skill you obtained during your previous training sessions to the real world.

Relaxing in a relaxed situation is great. However, as mentioned previously, you need Concentrated-Relaxation most during emergencies; after an accident; wet, cold, and in the middle of a tornado; during a hostile boardroom meeting; during grief and pain; after an emotional breakup; etc. Therefore, learn to mind-slide in real life situations, with open eyes, during distracting situations, in noisy environments, while watching videos, while stressed, etc.

Warning: Never ever use Concentrated-Relaxation or mind-sliding in situations where you have to be alert or when relaxation could become counterproductive or could cause accidents.

4.0 APPLICATIONS

4.1 Goal setting

Life is fair. If you ask for nothing, then that is exactly what you will get. Without a clear direction, there is no reason to move, because you will waste your time and resources by going around in circles.

Focus on maximizing your success. *"Success is to obtain what you seek and escape what you suffer"* [1] §62. However, what do you want to obtain? What do you want to escape? People often obtain what they seek, but then they do not like what they have obtained. Others escape the things that they need most. There can be no success without a mission, goal, or target. So, what is your mission statement? Determine what success means for you, because it is a personal experience. Ultimately, success is an affecter. Success is a conscious feeling that is created by your subconscious. A rich person could still feel poor, while a poor person could feel rich.

"A room filled with gold and jade
cannot be defended competently.
Admiring wealth and arrogance
brings personal loss and misfortune" [1] *§9.*

Are you chasing your own mission or someone else's? Let no other person determine your mission, because this is about your life and your success. Follow your heart and listen to your head!

Sometimes there is so much noise that we forget what is important and what is not. So, ask yourself questions. Could you write down in one sentence what you want? What would you do right now, if you would have only one month left to live? What would you do when you had only one day left? What would you do with your last hour? Whatever the answer, that might be something you should do right now. Would you ring your mother, an old friend, your banker? What would you like to be written on your grave? Who or what makes you happy?

Never let an urge to win a competition distract you from maximizing your success. Accumulating resources for the sake of accumulation will keep you busy, but are you going to use those resources?

"Do not value knowledge
and the people will not strive.
Do not admire goods that are difficult to obtain
and the people will not steal.
Do not display what is desirable
and the people will not revolt" [1] *§9.*

You live in a hyperdynamic reality, therefore, goal setting is an ongoing process. Dependent on the actual situation, your goal might be a long-term mission, while at other times it might be more productive to pursue short-term opportunities or escape immediate threats. Do you have your priorities right? Without survival there will be no further success. What are you going to do after you have achieved success?

4.2 Ideas - Accelerated thinking

Static thinkers are forced to adjust the situation to their rigid way of thinking. Therefore, they are associated with conflict, aggression, control, effort, and force. Despite their efforts, reality is hyperdynamic and sooner or later, the situation will inevitably change. At that moment, their previously productive nemoniks become counterproductive and they will fail.

Frederich Hegel (1770-1831) was a German philosopher who introduced a dialectic describing the progress of knowledge. During the first stage of his dialectic, a tested description of reality is accepted and is called the '*thesis*'. During the second stage, that thesis is challenged by a tested and conflicting description of reality that is called the '*antithesis*'. During the third stage, this conflict is resolved by merging the thesis and antithesis in a '*synthesis*'. The synthesis becomes the new thesis and the process repeats itself.

In accord with Hegel's dialectic, knowledge develops as a cyclic process that moves each turn closer towards the ultimate truth.

Hegel's dialectic is demonstrated by the progress of knowledge about the essence of light. Sir Isaac Newton (1642-1727) proposed that light was like a stream of tiny particles flying through space like cannonballs. This thesis was in accord with his law of motion. In contrast, Thomas Young (1773-1829) conducted the famous double-split experiment that supported the widely accepted antithesis that light moves through space like a wave. Both arguments were correct and, therefore, the evidence that light consist of particles could not reject the evidence that light consist of waves and vice versa. As a result, both groups got stuck in criticizing each other. That criticism maintained a counterproductive debate that inhibited scientific progress for decades.

In 1905, Albert Einstein accepted Thomas Young's antithesis, but returned also to Newton's thesis that light manifests itself as a stream of particles. In Hegelian terminology, Einstein synthesized Newton's thesis with Young's antithesis. He proposed that light moves through space and transparent matter both as a particle and a wave.

Hegel's dialectic implies that most knowledge is a process, rather than a state. In accord, reality shows that the volume of scientific research accelerates daily and, therefore, the journey is not finished yet. As long as Hegel's dialectic continues, we cannot claim to know the absolute truth. The rational truth is a static belief in a specific description of reality. We are on a journey and what we know is the last step on that journey, rather than the destiny. Einstein's synthesis concerning the essence of light might be just another thesis awaiting the next antithesis.

The nemonik-accelerator is based on Hegel's dialectic. It is a cognitive tool that could increase your speed of thinking. If you want to maximize your success, then you need the nemonik accelerator in order to adjust your way of thinking to the actual situation. The accelerator will increase the speed of your thinking by searching for the inevitable antithesis as soon as a thesis is accepted. In addition, search for the inevitable synthesis as soon as the antithesis is accepted.

Hence, foster disagreement during agreement and agreement during disagreement.

Nemonik thinkers have no need to maintain the status quo by force. They just adjust their thinking to the situation, like sailors adjusting their sails to the forever changing wind. Playing the Devil's advocate will reduce the counterproductive effects of groupthink and cognitive dissonance that is fostered by static thinking. Nemonik thinkers are goal-oriented, rather than competition-oriented. They focus on success, rather than winning. That allows them to foster freedom, alignment, compassion, allies, and win-win strategies.

4.3 Ideas – Cards

Ideas are the most precious treasures. One great idea could make you an instant billionaire. It could make you the happiest person on the planet. Therefore, write down any idea that comes up in your mind. Your genius uses reason that reason does not understand. Even if the idea sounds ridiculous at that moment to your conscious, it might be valuable later on. Any idea is a good idea, whether it is yours or not. So, have pen and paper ready.

Your mind is extremely complex. Your brainware comprises billions of neurons with forever changing connections and messages. You mindware is more complex than any computer. After all, the mind designed the computer. In addition, you live in a hyperdynamic reality that is continuously changing. Therefore, memory and thinking take time. Be patient. Sleep on important issues before you decide. Give you mind time to process the information. Your mind cannot be hurried without counterproductive repercussions. Waiting for solutions will create stress. Use Concentrated-Relaxation to manage such stress. Rash actions are likely to increase the problems.

Imagine that you have a problem, and you are stuck. You have run out of ideas. Writers block etc. Shuffle the deck of nemonik cards (Exercise 3.3). Lay one card on the table. Now, think about your problem while watching that nemonik. Your *Nemonik Associative Network* will focus your mind on the problem-solving tools associated with that particular nemonik. For starters, it is likely to return the memorized icon, definition, synonyms, SWOT analyses, questions, and the wisdom of proverbs associated with that particular nemonik. After a few minutes, replace the nemonik card with the next one. Repeat the procedure until you have screened all 21 nemoniks. Write all your ideas down without criticism.

4.4 Ideas - Brainstorm

Conscious Rational Thinking and subconscious Affectorial Thinking are mutually exclusive. Hence, the logic, reason, and criticism of the OBJECTIVE and COLLECTIVE mindmodes will inevitably inhibit the CREATIVE mindmode. Therefore, thinking should be separated in stages. During the first stage, the CREATIVE mindmode should generate creative affecters such as ideas, inventions, and intuitions. During the second stage, the OBJECTIVE and COLLECTIVE mindmodes should subject those ideas, inventions, and intuitions to the logic and reason of rational thinking.

An individual brainstorm session is a cognitive tool to generate creative affecters such as ideas, inventions, and intuitions. It is done with a silent conscious mind. Your conscious should be only an observer. Make sure that you cannot be disturbed. Prepare by writing down the mission of the session in a few words. Place that statement in a location that you can see easily. Then empty your mind with Concentrated-Relaxation and write down whatever comes up in your mind.

DO NOT CRITICIZE YOUR IDEAS

No matter how strange, just concentrate on writing down your thoughts as fast as possible. Try to switch off your critical conscious and let your hidden genius do what it does best. Do not try to control your thoughts. Just let it go. Relax.

4.5 Ideas - Inner-team

Visualization is a strong mental tool that could be used to create an inner team of imaginary experts. The stereotypes for the Objective and Collective mindmodes could be respectively a scientist and a judge. Furthermore, an inventor could symbolize the Creative mindmode, while a programmer could represent the Reactive mindmode. You could strengthen the images of your experts by giving them extraordinary features. For example, the scientist could have wild hair like Albert Einstein; the judge could wear a gown and a large white wig; the inventor could have a head as a glowing light bulb; and the programmer could carry a tablet. Alternatively, you could imagine people you admire.

Figure 14: Inner-team.

Use Concentrated-Relaxation to create semiconscious dominance. As illustrated in the following Figure 14, visualize a boardroom where you chair a meeting with your experts. First, take your time to explain mentally the actual situation to your experts in detail. Go over the details of the problem in your mind. The more they know, the better their advice will be. Imagine their responses.

The scientist and judge are both rational thinkers who will focus on the facts. If you have no facts, then they cannot help you. Rational thinkers will try to convince the meeting with reason and logic. They are critical and will not let emotions blind their reasoning. They search for alternative explanations, concealed facts, errors, inaccuracies, incompleteness, mistakes, etc. They follow their head, rather than their heart.

The scientist will consider the effects of the unchangeable natural laws on the situation. Is this case indeed independent of what anyone believes? Do we have enough data? Can we apply statistics? Can we test the hypothesis? What about an experiment? Are we able to make observations? Is the argument logical, valid, and reliable? The strength of the scientist is realism, while the weakness is a lack of compassion.

The judge will evaluate whether the changeable rules of the collective will affect the outcome. Could we make a rule? What is the collective opinion? Is there any tradition? Do the rules apply? Can the rules be reinterpreted? Could we enforce the rules? Could we change the rules? Is there a majority for either side? Are there any traditions or morals to be taken into account? The strength of the judge is creating group stability, while the weakness is stagnating the group with rules and regulations. Groupthink and peer pressure will maintain the rules and evoke stagnation.

The inventor and programmer are both affectorial thinkers who trust their affecters coming from their geniuses. Both experts are by definition nonrational because they receive neither sufficient facts nor rational explanations from their geniuses. Affectorial thinkers are noncritical and will urge the meeting to trust them. They will try to convince the meeting with self-confidence and charisma. Otherwise, they cannot use the affecters from their geniuses. They rely on their subconscious and, therefore, they are in touch with their intuitions and emotions. They follow their heart, rather than their head. They might point out that it is counterproductive to be factually correct, while feeling negatively about it.

The inventor will explore the unknown future and, therefore, moves objective, collective, and reactive thinkers out of their comfort zone. The inventor might use brainstorm and randomized information. Or put the problem upside down. The inventor might walk around the problem and make random remarks about the situation. In accord, the inventor is likely to propose unrealistic solutions. Nevertheless, do not interrupt the inventor with criticism. It will inhibit the creative process. The inventor is very useful if you have no facts or experience with the situation. However, the inventor needs sufficient time to digest information and develop ideas. The inventor's weakness is proposing sometimes unrealistic ideas.

The programmer writes the mental code for the mindsets that is based exclusively on experiences and skills obtained in the past. In accord, the programmer is always ready and will advise you to apply the prepared mindset. Act right now and rely on skills, emotions, feelings, common sense, heuristics, impulses, and reactions. The programmer's strength is accuracy and speed. Weaknesses are rash decisions, stagnation, and static thinking. The stagnation is maintained by cognitive dissonance.

If you want to maximize your success, then it is in your interest to manage the scientist, judge, inventor, and programmer as your inner-team. Motivate them to speak up by actively listening to them. Praise them, rather than criticize them. I wish you success with your imaginary friends. Remember, imagination might be incorrect, but it never lies.

4.6 Decision making

Use each of the 21 nemoniks to check your strategic position and that of your opponents. Is there any strengths, weakness, opportunity, or threat that you can use to your advantage? Could you turn your weakness into strength, or a threat into an opportunity? Alternatively, could you turn the strength of your opponents into weakness, or their opportunity into a threat? Could the questions and proverbs provide inspiration? What is the best and worst scenario? Hope for the best, but prepare for the worst.

Do not focus only on what you escape, but evaluate also on what you are going to obtain. Make sure that what you obtain is not worse than what you have right now. Do not jump from the pan in the fire. On the other hand, do not focus only on what you obtain, but also on what you might going to lose. The grass next door is always greener. Each action evokes a counterproductive reaction.

We live in an era that fosters extremes with nemoniks such as ADVANCE, ACCUMULATION, and ACTION. If there is any problem, then most people demand an instant decision. However, making the wrong decision is worse than making no decision at all. Waiting is delaying a decision until it is the most productive time for that decision. There is nothing wrong with that. Just relax.

Revealing your decision will create a reaction. Check whether those who try to force your decision are stakeholders. Who benefits? On the other hand, procrastination is not the same as waiting. Procrastination is a counterproductive delay. It is a thief of time and opportunities. A nemonik thinker might wait, but never procrastinate.

If you have the time, then check whether your information is correct. The snake concealed in the grass might be more dangerous than the roaring lion. Speaking inhibits listening. If you always speak, then you only repeat what you already know. Words are like arrows. Once released, they cannot be taken back. Be humble. Listen and learn.

The conscious ponders, but the subconscious decides. As long as the subconscious has not decided, it might bombard the conscious with stress. However, after the subconscious has made a decision, it sends affecters to the conscious to enforce that decision. Many decisions are made in order to avoid the stress of not making a decision. However, if you cannot make a decision, then there might be hidden reasons in your subconscious why you cannot make that decision.

Not making a decision is a decision. Your genius might decide that it is not the right time to make a decision. Subconsciously, you might not trust the stakeholders. You might sit on the fence and wait for an opportunity. You might be cautious, have insufficient information, or there is a smoke-screen of emotional energy in the air. Your conscious should respect that. Rather than forcing a decision, your conscious could use Concentrated-Relaxation in order to manage the stress of waiting for a subconscious decision.

4.7 Decision making - Coin

Modern technology shows that there are many physiological processes hidden outside our conscious awareness. That includes heartrate, brainwaves, skin temperature, etc. For example, biofeedback equipment aims to make us conscious aware of such processes by transforming those hidden signals into visible and audible ones. Biofeedback establishes a communication link between the conscious and subconscious. As a result, biofeedback participants can learn to control consciously their heartrate, brainwaves, skin temperature, etc.

As always, there is nothing new under the sun. Since the dawn of humanity, people have know the importance of this mental communication channel. Although the Ancients did not have our modern technology, they aimed for the same results with simpler tools such as cards, worry beads, bones, dices, coins, and pendulums.

In comparison with the conscious, the subconscious is omnipotent, because it creates the internal reality for the conscious. Therefore, it is no surprise that the Ancients believed that their conscious communicated with an external omnipotent entity, rather than their own subconscious. Modern technology shows us in real time what happens inside the subconscious. Similarly, the cards, bones, dices, coins, and pendulums provide information about our subconscious genius.

The subconscious is outside the conscious awareness. However, you could trick your reactive mindmode to send an affecter to your conscious. For that reason, you make a pseudo-decision. It does not matter what you decide. Just imagine that you have made up your mind and will execute your decision. As shown in the following Figure 15, take a coin and determine which decision is head (e.g. Yes) and which one is tail (e.g. No). Then flip the coin and watch your feelings carefully.

What comes up in your mind? The first feeling evoked by that fake decision might tell you what your subconscious thinks. Do you feel good about your imaginary decision? Unfortunately, the coin does not predict the future. It does not tell you what is correct or incorrect. However, it tells your conscious what you subconscious thinks right now. Intuition might be incorrect, but it never lies.

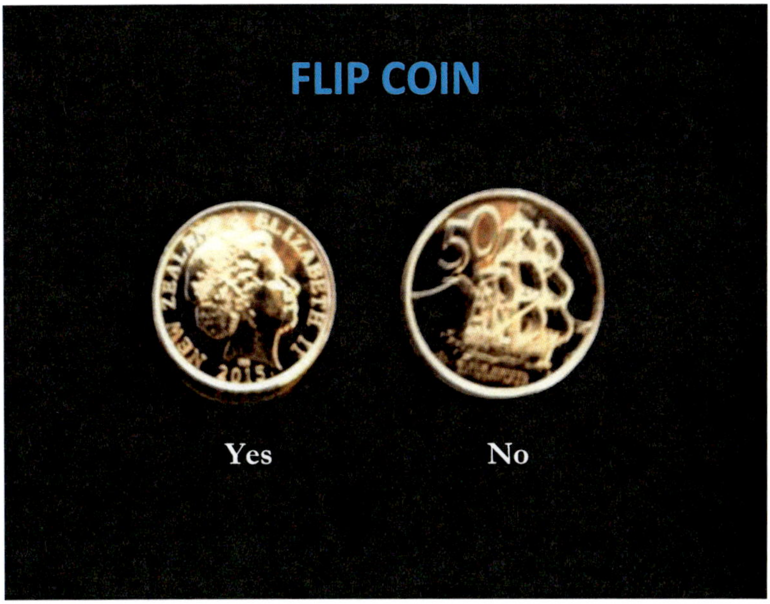

Figure 15: Flipping a coin.

4.8 Decision making - Pendulum

The justification for the use of tools such as cards, bones, dices, coins, and pendulums is provided in Application 4.8. As shown in the following Figure 16, a pendulum is a cognitive tool that comprises a small weight on a string. Just hold it as shown in the picture and relax your hand. While watching the pendulum, imagine the word 'YES'. Be patient, but your genius will either swing or rotate the pendulum in a particular direction. Do not try to affect the movement consciously. Just wait and see.

Subsequently, imagine the word 'NO'. Your genius will change the movement of the pendulum in another direction. Repeat the exercises until the responses are constant. At that point, your pendulum is ready to use. Watch it carefully, while imagining your real yes-no question. Do not try to influence the movement consciously. A pendulum might not give you the right answer, but it might tell you what your genius thinks. Intuition might be incorrect, but it never lies.

Figure 16: Pendulum.

4.9 Nemonik Information System (NIS) [11]

The Nemonik Information System (NIS) is an interactive thinking-tool providing information for the study and application of Nemonik Thinking (NT). NT provides a unique template for problem solving that comprises 21 keywords called nemoniks. Recalling a nemonik prompts the mind to evaluate a particular aspect of problem solving. Together, the nemoniks cover exhaustively all aspects including Lao Zi's principle of cosmic balance (4), the mind (4), reality (9), and the interaction between mind and reality (4). Using hyperlinks, the NIS provides the icon, definition, potential questions, SWOT analyses, mantras, and proverbs for that nemonik. Furthermore, the NIS comprises many useful features including an extensive glossary of nemonik terms and concepts. The NIS should be used in combination with the Nemonik Evaluation System (NES) [12].

4.10 Nemonik Evaluation System (NES) [12]

The Nemonik Evaluation System (NES) is an interactive tool for the evaluation of a person's knowledge about Nemonik Thinking (NT). NT is a unique template for problem solving that comprises 21 keywords called nemoniks. Recalling a nemonik prompts the mind to evaluate a particular aspect of problem solving. Together, the nemoniks cover exhaustively all aspects including Lao Zi's principle of cosmic balance (4), the mind (4), reality (9), and the interaction between mind and reality (4). Using hyperlinks, the NES evaluates your knowledge about the icon, definition, potential questions, SWOT analyses, mantras, and proverbs for that nemonik. The NES should be used in combination with the Nemonik Information System (NIS) [11].

5.0 COPING TOOLS

The mind is an associative cognitive system that jumps automatically from one concept to another. For example, recalling your kitchen in your mind might lead to a chain from dirty dishes, to last night's party, to a friend's problem. Sometimes, that associative process creates a counterproductive mental loop or compulsion. For example, Peter has no money, because he has no job— Peter has no job, because he has no car— Peter has no car because he has no money. Now, Peter is back at the beginning. That looping process is called 'worrying'. No matter how many times Peter will repeat to himself that he has no money that mantra will not improve his financial position. Instead, worrying is counterproductive, overloads the conscious, and inhibits rest and thinking. Peter needs to escape from his worry-loop and to restart his thinking.

No matter how much you worry, it will not solve anything. You will only lose energy and time. Therefore, you need coping strategies such as talking, mantras, private paradise, worry beads, burn letters, grief rituals, grief time, and bedside notes. They will restore the balances, within and between, the external and internal realities.

5.1 Talking

There might be so much going on in your mind that you withdraw inside. However, that is often also the time that you need the support of your friends and family most. Talking about unsolvable problems could help you to reorganize the frozen information in your mind. In addition, talking could increase the support from other people. If they do not know about your problems, then they cannot help you. Opening up can be extremely difficult, but simply talking about your problems will reset your own mind. As they say: *A shared problem is halve a problem.* A simple solution such as talking is often the best one. No matter how high the pressure, keep the external communications open. They are your lifelines. A hug can be much more productive than a cold shoulder.

5.2 Mantras

Your subconscious decides how your conscious feels. However, your conscious is the mind's manager. Therefore, you are in control. If any feeling becomes counterproductive, you could manage it consciously with mantras. I do not know about you, but I cannot feel sad and happy at the same time. You can use that to your advantage. As illustrated in the following Table 15, create a mantra that evokes a mutually exclusive feeling. For example, you might use [>I feel< | <great>] or [>I feel< | <happy>]. Those mantras are incompatible with sadness.

Mutually exclusive mantras.
Stage 2: Evoking semiconscious dominance
Stage 7: Mutually exclusive mantra
e.g. >I feel< | <great>
Stage 6: Returning conscious dominance

Table 15: Mutually exclusive mantras.

A mantra is not a declaration of your current state of mind. Instead, it is an conscious instruction to your subconscious to foster that state of mind. The mantra will not be true in the beginning, but fake it until you make it. If you are persistent, then your genius will change the affecters to your conscious. All things will pass.

5.3 Nemonik Bracelet

The subconscious is a multitasker that can execute several tasks at once. On the other hand, the conscious can only execute one task at the time. You could use that limitation of your conscious to your advantage.

Since ancient times, strings of beads have been used as mnemonics or memory aids like the Dominican Rosary (50 beads) and the Japamala of the Hindus (108 beads). What has worked for thousands of years for many different cultures must have some intrinsic value. Without relying on spiritual, religious, or paranormal entities, nemonik thinkers use a string of beads as a tool to prompt their memory and thinking, while reducing their stress.

As shown in Figure 17, the nemonik bracelet comprises 21 beads. Each bead represents a specific nemonik. They are divided into six colour sections: (cosmic principle-4 beads) (mind-4 beads) (space-3 beads) (matter-3 beads) (time-3 beads) and (interaction-4 beads). The size, shape, and colours of the beads have no special meaning. Be creative and take any beads you like as long as they are different for each section.

The aim of the nemonik bracelet is to change unproductive worrying into productive thinking. Each bead of the bracelet recalls a specific nemonik that activates subconscious problem-solving tools. Starting with the focus-bead, one goes through the beads one by one.

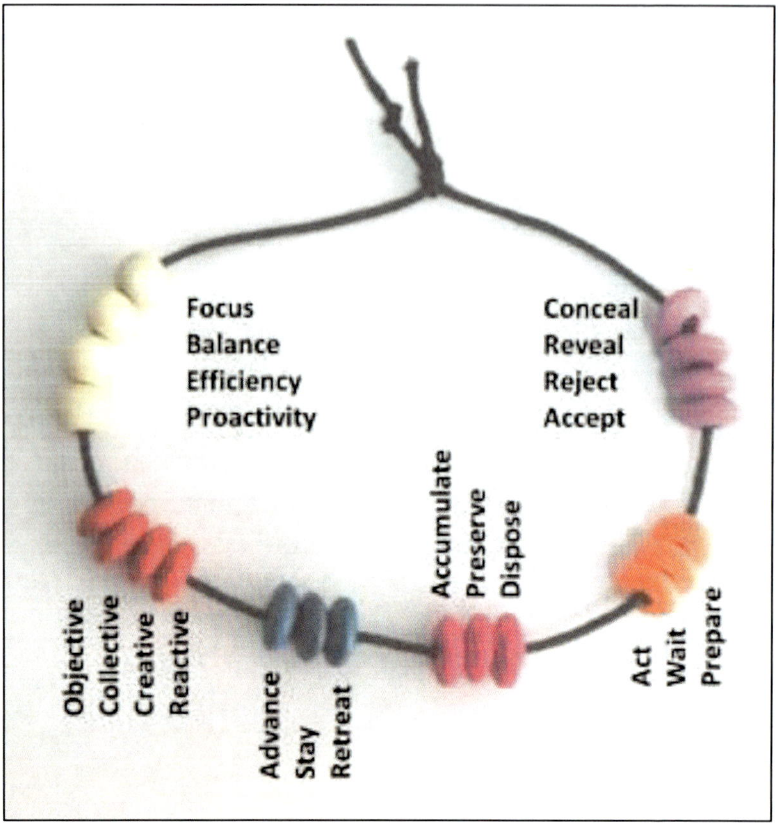

Figure 17: Nemonik bracelet.

Worries can be persistent. Do not despair if the worry-loop is not broken immediately. Be patient. Increase the number of Stress-inhibitor loops for each round with one, until the worry-loop remains silent: (1st time-1 loop) (2nd time-2 loops) etc. Once started, always finish the particular round. Otherwise, you might reinforce yourself to give up. Whenever, the worry-loop returns in the mind, as it often does, just apply the bracelet immediately. The nemonik bracelet inhibits worrying by occupying the conscious and, therefore, it creates peace of mind. It turns the time spend on unproductive worrying into productive problem-solving, while reducing your stress.

5.4 Private Paradise

The subconscious might pressure the conscious to take productive actions that maximize success. However, sometimes there is a conflict between the external and internal realities that cannot be solved immediately. As a result, the combined external and internal pressures on your conscious could become counterproductive. The mind could become a pressure cooker in which the pressure itself prevents a release of that pressure. Ultimately, that increasing pressure could lead to counterproductive anxiety, depression, anger, etc.

You might not be able to reduce the external pressure. However, your conscious is the manager of your subconscious, while your subconscious controls your conscious perception of that reality. Hence, your conscious manager could lower the internal pressure. If you mind functions again, then you can use that break to find a solution for the external problem.

This exercise will help your conscious manager to lower temporarily the internal pressure. This application needs several sessions of preparation. As illustrated in the following Table 16, use Concentrated-Relaxation to evoke semiconscious dominance.

Creating a Private Paradise.
Stage 2: Evoking semiconscious dominance
Stage 8: Imagination
Imagine private paradise [A]
Stage 6: Returning conscious dominance

Table 16: Private Paradise

Imagine in your mind a peaceful place where you can withdraw from the turmoil of life. Do not worry. If you cannot visualize it, then just think about it. A private paradise. Imagine a favourite and peaceful location. It might be either real or fantasy. As shown in the following Figure 18, you could imagine a peaceful pond. Whatever you select, it is recommended to train yourself in visiting your paradise before you have any problems.

During each subsequent session add more details to your paradise. That might be imaginary items, colours, fragrances, sounds, etc. Whatever makes you feel great and relaxed.

If you have practised this exercise enough, then you can evoke you private paradise when you need a break. Remember, this is not a permanent solution. It gives you a break to find one.

Figure 18: Private paradise.

5.5 Burn Letters

Writing letters about unsolvable problems could help to reorganize the information in your mind. Often, losing people leaves things unresolved and unsaid. Some people find relieve in writing about their feelings. Write down whatever you would like to have said. Burning such letters might bring a kind of closure.

5.6 Grief sessions

Emotional and material losses create often unsolvable problems. To remain functional, it might help to set a daily time to grief. During the day, the conscious could promise the subconscious a grief session at a set time. For example, your conscious makes a promise to your subconscious to grief from 19.00 to 20.00 hrs. Each time the stress increases during the day, repeat the promise to grief at the set time. In that way, you remain functional during the rest of the day. Take the agreed time to grief. Let yourself go. Talk to yourself and cry if you feel so. Give you subconscious an opportunity to express its loss. Keep going until the end of the session even if the grief stopped. Stop the grieving exactly at the end of the session. If still grieving, just promise a new grief session to your subconscious. Keep each conscious promise. Never cheat on your genius. Otherwise, the strength of this technique is likely to weaken.

5.7 Grief rituals

Grief rituals could help to deal with unsolvable problems caused by emotional or material losses. Often, people do not get the opportunity to say a last goodbye. To recreate the feeling associated with such farewell, a grieving person could let a balloon fly away, throw a bottle with a message in the sea, bury a meaningful item, burn a candle, etc. In every culture, rituals serve to deal with the grief of the survivors.

5.8 Bedside Notes

Some people worry so much at night that they wake up. In that case, it might help to have pen and paper on a bedside table. Writing down the idea or problem might relax your subconscious, because the topic cannot be forgotten. You can go back to sleep, because the problem has been taken care off.

6.0 ANCIENT SCIENCE

6.1 Lao Zi's Dao De Jing

老子之 道德經

The Way of Nature

and

The Way of People

Dr Auke Schade

道

Dao - The Way of Nature

Chapters §1—37

§ 1

道可道也非恒道也.　名可名也非恒名也.　无名万物之始也.　有名万物之母也.　故恒无欲也以观其眇.　恒有欲也以观其曒.　此两者同出.　异名同谓.　玄之又玄，众眇之门.

The Way that can be named is not the eternal Way. The name that can be named is not the eternal name. Nameless, it is the origin of All-things. Being named, it is the Mother of All-things. Therefore, be always without desire and see the details. Have always desires and see the limits. These two things occur together. Different names with the same meaning. Profoundly mysterious, they are the gateway to many details.

§ 2

天下皆知美之为美也斯恶已.　皆知善之为善，斯不善已矣.　故，有无之相生也.　难易之相成也.　长短之相刑也.　高下之相盈也.　音声之相和也.　先后之相随恒也.　是以圣人処无为之事.　行不言之教.　万物作而不始也.　为而弗恃也.　成功而弗居也.　夫唯弗居也是以弗去也.

If everyone in the world recognises beauty as beautiful, then there is already ugliness. If everyone recognizes good as goodness, then there is already badness. Therefore, Existence and Non-existence generate each other. Difficult and easy turn into each other. Long and short shape each other. High and low fill each other. Tone and voice harmonise each other. Before and after follow each other forever. Therefore, sages manage their affairs with Non-action. They carry out their teachings without speaking. All-things rise, but do not initiate them. Act, but do not rely on it. Succeed, but do not claim. Only that what is not claimed can therefore not be taken away.

§ 3

不上贤使民不争. 不贵难得之货使民不为盗. 不见可欲使民心不乱. 是以，圣人之治也虚其心. 实其腹弱其志，强其骨. 恒使民无知无欲也. 使夫之知者不敢为而已也. 为无为则无不治矣.

Do not promote the knowledgeable and the people will not strive. Do not admire goods that are difficult to obtain and the people will not steal. Do not display what is desirable and the people will not revolt. Therefore, sages rule by emptying the minds of people, filling their stomachs, weakening their ambitions, and strengthening their bones. Let the people be always without knowledge and without desire. Let those who know, not dare to act but stop. Act with Non-action, then there will be no anarchy.

§ 4

道冲，而用之有弗盈也. 渊呵！ 似万物之宗；锉其兑, 解其纷, 和其光，同其尘. 湛呵. 似，若存. 吾不知其谁之子也，象帝之先.

The Way is empty, but use it and it has not to be refilled. It is so deep! Like the ancestor of All-things: it smooths their blending, untangles their disorder, softens their glare, and merges their dust. Invisible. Nevertheless, it seems to exist. I do not know whose child it is, but it seems to predate the Emperor.

§ 5

天地不仁，以万物为刍狗.　圣人不仁，以百姓为刍狗. 天地之间其犹橐籥与.　虚而不淈.　动而俞出.　多闻数穷不若守于中.

The Sky and the Earth are not benevolent, because All-things act as straw dogs. Sages are not benevolent, because common people act as straw dogs. What is between the Sky and the Earth is like a bagpipe. It is empty, but not exhausted. Use it and more will be produced. Listening to many details is exhausting and not as good as following your heart.

§ 6

谷神不死是谓玄牝.　玄牝之门是谓天地之根.　绵绵其若存用之不勤.

The immortal Valley Spirit is called the Mysterious Female. The home of this Mysterious Female is called the origin of the Sky and the Earth. It seems to exist forever and using it is no hard work.

§ 7

天长地久.　天地之所以能长且久者?　以其不自生也.
故，能长生.　是以，圣人退其身.　而身先.　外其身.
而身存.　不以其无私欤，故，能成其私.

The Sky endures and the Earth last long. Why do the Sky and the Earth last long and endure? That is, because they do not foster themselves. Therefore, they can live long. Accordingly, sages withdraw themselves. Yet they are first. They put themselves outside. Yet they are inside. Because they are selfless, therefore, their self-interest is fulfilled.

§ 8

上善.　如水水善利万物而不争.　处众人之所恶. 故,
几于道矣.　居善地.　心善渊.　予善仁.　言善信.
正善治.　事善能.　动善时. 夫唯不争故无尤矣.

Superior goodness is like water. The goodness of water benefits All-things and it does not strive. It occupies places everybody dislikes. Therefore, it is close to the Way. In dwelling, the goodness is location. In thinking, the goodness is depth. In giving, the goodness is benevolence. In speaking, the goodness is truth. In ruling, the goodness is order. In working, the goodness is skill. In action, the goodness is timing. Only those who do not strive will therefore not fail.

§ 9

持而盈之不若其已. 揣而锐之不可长保之也. 金玉盈室莫之能守也. 贵富而骄自遗咎也. 功遂，身退天之道也.

Accumulating and filling up is not as good as stopping in time. Hammer it too sharp and it cannot last long. A room filled with gold and jade cannot be defended competently. Admiring wealth and arrogance brings personal loss and misfortune. When merit is achieved, withdrawing yourself is the Way of nature.

§ 10

载营魄抱一能*毋*离乎. 专气至柔能如婴儿乎. 修除玄. 监能*毋*有疵乎. 爱民治国能*毋*以知乎. 天门开阖能为雌乎. 明白四达能*毋*以知乎. 生之畜之. 生而弗有. 长而弗宰也. 是谓玄德.

Carry the *'corps de esprit'* and unite it inseparable with the One. Concentrate vital energy and be as soft as an infant. Study and eliminate mysteries. Inspect them competently without flaws. Love the people and rule the country without using knowledge. Open and close the gates of nature as a female. Understand the surroundings without using knowledge. Generate them and raise them. Generate but do not possess. Develop but do not exploit. This is called profound virtue.

§ 11

三十辐同一毂，当其无有车之用也. 捻埴以为器，当其无有埴器之用也. 凿户牖以为室，当其无有室之用也. 故，有之以为利，无之以为用.

Thirty spokes merge into one hub, but its Non-existence is useful for a carriage. Moulded clay makes a cup, but the Non-existence of clay is useful for a cup. Chiselled doors and windows make a room, but their Non-existence is useful for a room. Therefore, using Existence is beneficial, while using Non-existence is useful.

§ 12

五色使人目盲. 五音使人之耳聋. 五味使人之口爽. 驰骋田猎使人心发狂. 难得之货使人之行妨. 是以圣人之治为腹而不为目. 故，去罢耳此.

The five colours will blind people's eyes. The five tones will deafen people's ears. The five flavours will refresh people's mouth. Galloping and hunting in the field will overexcite people's minds. Goods that are difficult to obtain will harm people. Therefore, sages will rule for the stomach and not for the eyes. Therefore, reject that and accept this.

§ 13

宠辱若惊. 贵大患若身. 何谓宠辱若惊? 宠之为下也. 得之若惊失之若惊. 是谓宠，辱若惊. 何谓贵大患若身？ 吾所以有大患？ 者，为吾有身也. 及吾无身，吾有何患乎？ 故，贵以身于为天下若可以寄于天下矣. 爱以身为天下者可以寄天下矣.

Favour and disgrace are just like distress. They cause great suffering just like the body. Why saying that favour and disgrace are just like distress? Favour is inferior. Receiving it is like distress and losing it is like distress. Whether it is called favour or disgrace, it is like distress. Why saying that it cost great suffering just like the body? Why do I have great suffering? That is, because I have a body. If I had no body, how could I suffer? Therefore, those who purposely value their body for serving the world can be entrusted with the purpose of the world. Those who purposely love their body for the world can be entrusted with the world.

§ 14

视之，而弗见. 名之曰微. 听之，而弗闻. 名之，曰希. 摶之而弗得. 名之曰微. 此三者不可致计, 故混而为一，一者其上不谬. 其下之不忽. 寻寻呵. 不可名也. 复归于无物. 是谓无状. 之状无物之象. 是谓惚恍. 迎之，而不见其首. 随之，而不见其后. 执今之道以御今之有. 以知古始是谓道纪.

Look at it, yet it cannot be seen. Its name is called invisible. Listen to it, yet it cannot be heard. Its name is called inaudible. Seize it, yet it cannot be caught. Its name is called insubstantial. These three phenomena cannot be extensively evaluated, because they merge into the One. Above the One there is no void. Below it there is no substance. It is infinite. It cannot be named. Every time it returns to nothing. It is called shapeless. Like the shape of nothing. It is called dim and elusive. Face it, yet you do not see its head. Follow it, yet you do not see its back. Adhere to the present Way in order to manage the present Existence. Use it to understand its ancient origin. That is called the principle of the Way.

§ 15

古之善为道者． 微眇玄达． 深不可识．夫唯不可识．故，强为之容曰：“与呵，其若冬涉川． 犹呵，其如畏四邻．俨呵，其若客． 涣呵，其若凌泽．沌呵，其若朴． 旷呵，其若浴． 湷呵，其若浊”．浊以静之者将徐清． 安以动之而者将徐生． 葆此道者不欲盈． 夫唯不欲盈是以能敝不成．

Those Ancients who practised the Way competently, understood profoundly the smallest details. Their depth cannot be known. They cannot be understood. Therefore, they are difficult to describe and called: *"Careful, like they were wading through a river in the winter. Hesitant, like they were afraid of their surrounding neighbours. Solemn, like they were guests. Dissipating, like they were melting snow. Vague, like they were simple. Empty, like they were a valley. Merging, like they were mud"*. Do not stir mud and it will slowly clear. If settled, stir it and then it will slowly come alive. Those who keep the Way do not desire fullness. Only those who desire no fullness are therefore able to exhaust themselves without renewal.

§ 16

致虚极也.　守静笃也.　万物傍作，吾以观其复也.

夫物秐秐各复归于其根.　归根曰静.　静是谓复命.　复命常也.　知常明也.　不知常妄.　妄作凶.　知常容.　容乃公.　公乃王.　王乃天.　天乃道.　道乃久.　沕身不殆.

Concentrate on removing extremes. Nurse tranquillity faithfully. All-things around us rise, and I watch them return. Those things are numerous and each one returns to its roots. Returning to the roots is called tranquillity. Tranquillity is called returning to order. Returning to order is a constant. Knowing this constant is brilliant. Not acknowledging this constant is arrogant. Arrogance causes misfortune. Knowing this constant is embracing. Embracing is honourable. Honourable is Kingly. Kingly is natural. Natural is the Way. The Way is forever. It provides a content life without danger.

§ 17

大上下知有之.　其次亲誉之.　其次畏之.　其次其下侮之.　信不足，焉有不信.　猷呵其贵言也成功遂事.　而，百姓谓；我自然也.

Great leaders are those known by their subjects to exist. Next are those that are loved and praised. Next are those that are feared. Next are those low ones who are insulting. If there is not enough trust, then there is distrust. Those of value speak about their plans and succeed in completing their affairs. Yet, the common people will say; it happened naturally.

§ 18

故，大道废，焉有仁义． 知快出焉有大伪． 六亲不和焉有畜兹． 国家昏乱焉有贞臣．

Therefore, if the great Way is rejected, then there will be benevolence and justice. Knowledge and cleverness will appear and then there is great hypocrisy. Family relationships will be disharmonious and then there is animal dirt everywhere. The State's household will be a confused disorder and then there is bureaucracy.

§ 19

绝圣弃知而民利百倍． 绝仁弃义而民复孝慈． 绝巧弃利盗贼无有． 此三言也以为文而未足也． 故，令之有所属． 见素抱朴． 少私而寡欲．

Discard adoration and reject knowledge and the people will benefit a hundred times. Discard benevolence and reject righteousness and the people will return to filial piety and compassion. Discard cleverness and reject profit and there will be no burglars and thieves. These three declarations could be regarded to be inadequate slogans. Therefore, let the people have also institutions. Show modesty and embrace simplicity. Lack selfishness and restrain desires.

§ 20

绝学无忧. 唯之与诃: 其相去几何? 美与恶: 其相去何若? 人之所畏, 亦不可以不畏人. 人望. 呵其未央哉. 众人熙熙. 若乡于大牢而春登台. 我泊兮未兆. 若婴儿未咳. 儽呵; 似无所归. 众人有余. 而我独若遗. 我愚人之心也惷惷兮. 俗人昭昭. 我独若昏呵. 俗人察察. 我独闷闷呵. 淡呵. 其若海朢呵. 其若无所止. 众人皆有以. 我独顽以悝. 吾欲独异于人而贵食母.

Discard knowledge and there are no worries. Flattery and rebuke: how much do they differ from each other? Satisfaction and dissatisfaction: how much do they differ from each other? What everybody fears, one has to fear as well. Everybody stares at me. They do not stop. Everybody is very happy. Just like a big sacrificial feast in the village and stepping on stage in the springtime. I am quiet and not predictable. Just like a baby that has not coughed yet. Tired; without a place to return to. Everybody has a surplus. Yet, only I seem to be lacking. I am a very stupid fool in other people's minds. Everybody is very clear. Only I seem to be confused. Everybody is very certain. Only I seem to be very uncertain. They are indifferent. It is like staring at the sea. It is like having no place to rest. Everybody has a purpose. Only I am stubborn and my purpose seems to be ridiculous. Only I desire to differ from other people and value the food from the Mother.

§ 21

孔德之容唯道是从. 道之物唯恍唯惚. 惚呵. 恍呵. 中有象呵. 恍呵. 惚呵. 中有物呵. 幽呵. 冥呵. 其中有精呵. 其精什真. 其中，有信. 自古及今，其名不去. 以，顺众父. 吾何以知众父之然也？ 以此.

The greatest virtue is following only the Way. The contents of the Way are only elusive and dim. Dim. Elusive. Inside there are images. Elusive. Dim. Inside there are things. Tranquil. Obscure. Its centre has energy. Its energy is very real. Inside it, there is information. From past to present times, its name was never erased. Therefore, align with the Father of the Multitude. How do I know that the Father of the multitude is like this? From this account.

§ 22

曲则全. 枉则正. 洼则盈. 敝则新. 少则得. 多则惑. 是以，圣人执一. 以为, 天下牧. 不自见也故明. 不自视故章. 不自伐故有功. 弗矜. 故，能长. 夫唯不争. 故，天下莫能与之争. 古之所谓："曲则全者". 几语也？ 才，诚全归之.

Bend then be preserved. Twist then be straightened. Empty then be filled. Exhaust then be renewed. Lack then receive. Have surplus then be confused. Therefore, sages hold on to the One. Accordingly, they are the shepherds of the world. They do not display themselves. Therefore, they are brilliant.

They do not regard themselves. Therefore, they are honoured. They do not boast about themselves. Therefore, they have merit. They are not arrogant. Therefore, they will develop. They do not strive. Therefore, no one can strive with them. The ones called the *'Ancients'* said: *"Those who bend will be preserved"*. Is that saying insignificant? However, true preservation was their return.

§ 23

希言自然. 飘风不冬朝. 暴雨不冬日. 孰为此? 天地. 而，弗能久. 有兄人于乎? 故，从事而道者同于道. 从事于德者同于德. 从事于失者同于失. 同于德者道亦德之. 同于失道亦失之.

Speaking seldom is natural. Violent storms do not drum all morning. Violent rains do not drum all day. Who serves them? The Sky and the Earth. Yet, they cannot go on forever. So how about people? Therefore, those who submit their affairs to the Way will merge with the Way. Those who submit their affairs to virtue will merge with virtue. Those who submit their affairs to loss will merge with loss. Those who merge with virtue will also gain the Way. Those who merge with loss will also lose the Way.

§ 24

跂者不立. 自见者不明. 自是者不章. 自伐者无功. 自矜者不长. 其在道也曰: “余食赘行”. 物或恶之. 故，有欲者弗处也.

Those who stand on tiptoe do not stand firm. Those who display themselves are without brilliance. Those who regard themselves are without honour. Those who boast about themselves are without merit. Those who are arrogant are without development. Their Way is called: *"Leftover food and unnecessary action"*. These things are disgusting. Therefore, those who have desires will not succeed.

§ 25

有物昆成先天地生. 萧呵. 寥呵. 独立而不改. 可以为天地母. 吾未知其名字之，曰道. 吾强为之名，曰大. 大曰逝. 逝曰远. 远曰反. 故道大；天大；地大；王亦大. 国中有四大而王居一. 焉，人法地；地法天；天法道；道法自然.

There was a thing undivided and complete before the Sky and the Earth were born. Desolate. Empty. Independent and unchanging. Yet, it acts as the origin of the world. I do not know how its name is pronounced, but I call it the Way. If I were forced to describe it, then I would call it great. Great means continuous. Continuous means forever. Forever means returning. Therefore, the Way is great. The Sky is great. The Earth is great. The King is also great. Inside the Universe there are four Greatnesses and the King is one. Therefore, people follow the Earth, the Earth follows the Sky, the Sky follows the Way, and the Way follows nature.

§ 26

重为轻根.　静为躁君.　是以，君子终日行不离其辎重.　虽, 有环馆燕处则, 超然.　若何万乘之主而以身轻于天下.　轻则失本.　躁则失君.

Heaviness is the foundation of lightness. Tranquillity is the sovereign of rashness. Therefore, great men who travel all day will not leave their heavy wagons. Although, there is a walled guest house in a quiet place nearby, they remain aloof. Just like a lord with ten thousand chariots who considers himself less important than the State. Lightness will lose the foundation. Rashness will lose the sovereign.

§ 27

善行者无辙迹.　善言者无瑕适.　善数者不用筹策.　善闭者无关籥.　而，不可启也.　善结者无绳约.　而，不可解也.　是以，圣人恒善救人而无弃人.　物无弃财.　是谓袭明.　故，善人者不善人之师.　不善人善人之资也.　不贵其师，不爱其资，虽知乎，大迷.　是谓眇要.

Competent travellers leave no trail. Competent speakers pursue no flaws. Competent accountants do not use bamboo counting sticks. Competent wardens lock without keys. Yet, it cannot be opened. Competent weavers arrange without strings. Yet, it cannot be untied. Therefore, sages save people always competently by not rejecting them. Not rejected things are resources. Accordingly, that is called brilliant. Therefore, competent people are the teachers of incompetent people. Incompetent people are the resources of competent people. Those who do not value their teachers, or do not love their resources, although knowledgeable, are greatly confused. This is called the essential detail.

§ 28

知其雄守其雌为天下溪. 为天下溪恒德不离. 恒德不离，复归于婴儿. 知其白守其黑为天下式. 为天下式恒德不忒. 恒德不忒，复归于无极. 知其白守其辱为天下浴. 为天下浴恒德乃足. 恒德乃足，复归于朴. 朴散则为器. 圣人用之则为官长. 夫大制无割.

Know the male and guard the female and become the stream of the world. Be the stream of the world and the eternal virtue never leaves. If the eternal virtue never leaves, then you will return to infancy. Know the white and guard the black and be the example for the world. Be the example for the world and the eternal virtue never errs. If the eternal virtue never errs, then it returns to moderation. Know the pure and guard the disgrace and be the valley of the world. Be the valley of the world and the eternal virtue will always be enough. If the eternal virtue is always enough, then you will return to simplicity. Simplicity breaks up and then it becomes tools. Sages use them and become official leaders. They will not divide great organisations.

§ 29

将欲取天下而为之，吾见其不得已.　天下神器也非可为者也.　为之者败之.　执之者失之.　物或行或隋.　或热或坐.　或强或羸.　或培或堕.　是以圣人，去什，去大，去诸.

If anyone would desire to take the world and interfere with it, I see that they have no alternative. The world is a container of energy that cannot be interfered with. Those who interfere will fail. Those who hold will lose. Things may succeed or may fail. They may be hot or may be cold. They may be strong or may be weak. They may grow or may decay. Therefore, sages reject extremes, reject grandeur, and reject extravagance.

§ 30

以道佐人主者.　不欲以兵强于天下.　其事好还. 师之所居楚朸生之.　善者果而已矣.　毋敢以取强焉.　果而弗矜.　果而弗伐.　果而毋骄. 果而毋得已.　居是谓果而不强. 物壮而老.　是谓之不道.　不道早已.

Use the Way to assist the leaders of people. Do not use soldiers to force the world. Such actions are likely to rebound. Where armies have camped only thorny bushes will grow. Those who are competent succeed and stop in time. Do not dare to take power. Succeed without boasting. Succeed without attacking. Succeed without arrogance. Succeed without excess. That is called succeeding without force. Strong things will become weak. They are not called the Way. What is not the Way will soon perish.

§ 31

夫兵者不祥之器也. 物或恶之. 故，有道者弗居. 君子居则贵左. 用兵则贵右. 兵者不祥之器也. 故，兵者非君子之器也. 不得已而用之. 铦袭为上，勿美也. 若美之，是乐杀人也. 夫乐杀人不可以得志于天下矣. 是以，吉事上左. 丧事，上右. 是以，便将军居左，而上将军居右. 言以丧礼居之也. 杀人之，众以哀悲莅之. 战胜而以丧礼处之.

Armies are the tools of misfortune. They are disgusting. Therefore, those who possess the Way will not claim them. Great men will occupy and value the unorthodox. They will use soldiers who value the orthodox. Armies are the tools of misfortune. Therefore, armies are not the tools of great men. When there is no alternative then use them. Attack with sharp weapons to become victorious, but use them without satisfaction. Those who are satisfied by it, like to kill people. Those who like to kill people cannot achieve the goals of the State. Therefore, during fortunate events, the left side is honoured. During funerals, the right side is honoured. Therefore, junior generals occupy the left side, while senior generals occupy the right side. Their places are determined in accord with funeral ceremonies. If people were killed, then many will attend with sadness. Hence, treat battle victories as funeral ceremonies.

§ 32

道恒无名. 朴唯. 而, 天下弗敢臣也. 侯王若能守之, 万物将自宾. 天地相合也, 以降甘露.
民莫之令, 而自均焉. 始制, 有名. 名亦既有, 夫亦将知止. 知止所以不殆. 譬道之在天下也, 犹川浴之, 与江海也.

The Way is forever nameless. It is so simple. Yet, the world should not dare to control it. If Marquises and Kings would follow it, then All-things would submit themselves. If the Sky and the Earth would unite with each other, then it would rain sweet dew. People would not have to be ordered, but they would balance themselves. If an organisation is established names will appear. If names appear, then know that it is time to stop. Know when to stop and there will be no danger. The Way is to the world, what a valley is to a river, and what a river is to the sea.

§ 33

知人者智也. 自知者明也. 胜人者有力也. 自胜者强也. 知足者富也. 强行者有志也. 不失其所者久也. 死, 而不忘者, 寿也.

Those who know other people are wise. Those who know themselves are brilliant. Those who overcome other people have power. Those who overcome themselves are strong. Those who know what is enough are rich. Those who are strong pioneers have ambition. Those who do not lose their institutions will last long. Those who die, but are not forgotten, will live on.

§ 34

道渢呵．　其可左右也．　成功遂之事．　而，弗名有也．万物归焉．　而，弗为主．　则恒无欲也．　可名于小矣．万物归焉．　而，弗为主．　可名于大矣．　是以圣人能成大也，以其不为大也．　故，能成其大．

The Way floats. It can be unorthodox or orthodox. It completes its affairs successfully. Yet, it is not a famous being. All-things return to it. Yet, it does not act as their master. It is always without desire. Hence, it could be named small. All-things return to it. Yet, it does not act as their master. Hence, it could be named great. Therefore, sages can achieve greatness, because they do not act great. Therefore, they can achieve greatness.

§ 35

执大象天下往．　往而不害平大．　乐与饵过客止．　故，道之出言也曰：淡呵其无味也．　视之，不足见也．　听之，不足闻也．　用之，不可既也．

Hold on to the great image, and the world will come. It will come without harm and with great calm. Music and food will stop passing travellers. However, words describing the Way are called: *bland and without taste.* Look at it, and there is not enough to see. Listen to it, and there is not enough to hear. However, use it and it cannot be depleted.

§ 36

将欲翕之，必古张之.　将欲弱之，必古强之.　将欲去之，必古与之.　将欲夺之，必古予之.　是谓微明. 柔弱胜强. 鱼不可脱于渊.　邦之利器不可以示人.

To fold something, it must have been unfolded before. To weaken something, it must have been strengthened before. To abandon something, it must have been attached before. To seize something, it must have been separated before. This is called profound brilliance. The soft and weak will overcome the strong. A fish should not leave the deep water. The sharp weapons of the State should not be used in view of the people.

§ 37

道恒无名也.　侯王若能守之，而万物将自化.　化而欲作，吾将镇之以无名之朴.　镇之以无名之朴夫将不辱. 不辱以静天地将自正.

The Way is forever nameless. If Marquises and Kings could follow it, then All-things would transform themselves. If this transformation would cause desire, then I would suppress it by using the simplicity of the nameless. Suppressing it by using the simplicity of the nameless will not disgrace them. Use tranquillity without disgrace and the world will regulate itself.

德

De - The Way of People

Chapters §38—81

§ 38

上德不德.　是以，有德.　下德不失德.　是以，无德. 上德无为而无以为也.　上仁为之而无以为也.　上义为之而有以为也.　上礼为之而莫之应也，则攘臂而乃之. 故，失道，而后德，失德，而后仁.　失仁，而后义. 失义，而后礼.　夫礼者忠信之泊也，而乱之首也.　前识者道之华也而愚之首也.　是以，大丈夫居其厚而不居其泊.　居其实而不居其华.　故，去彼而取此.

Superior virtue pursues no virtue. Therefore, it is virtue. Inferior virtue pursues virtue. Therefore, it is no virtue. Superior virtue uses Non-action and there is no action used. Superior benevolence acts and yet there is no use in those actions. Superior justice acts and there is purpose in those actions. Superior propriety acts and if there is no agreement, then the arms are bared. Therefore, after the Way is lost there will be virtue. After virtue is lost, there will be benevolence. After benevolence is lost, there will be justice. After justice is lost, there will be propriety. Those who have propriety possess only a thin layer of loyalty and sincerity, which is the beginning of disorder. Those who pretend to know the future are the fruitless flowers of the Way and the chiefs of fools. Therefore, great men occupy the thick and do not occupy the thin. Occupy the fruit, but do not occupy the fruitless flowers. Therefore, reject that and accept this.

§ 39

昔之得一者：天得一以清；地得一以宁；神得一以灵；浴得一以盈；侯王得一而以为天下正．　其至之一也：天毋已清，将恐裂；地毋已宁，将恐发; 神毋已灵，将恐歇；谷毋已盈；将将恐竭．　侯王毋已贵以高，将恐蹶．　故，必贵以贱为本．　必高而以下为基．　夫是以，侯王自谓不谷　‘孤寡’．　此其以贱之本与非也．　故，致数誉无誉．　是故，不欲琭琭若玉，落落若石．

Of those in the past that obtained the One: the Sky obtained the One through pureness; the Earth obtained the One through quietness; the mind obtained the One through effectiveness; the valley obtained the One through filling; and Marquises and Kings obtained the One by regulating the world. The conclusion about the One is: if the Sky is not clear yet, then fear that it will crack; if the Earth is not quiet yet, then fear that it will burst; if the mind is not effective yet, then fear that it will cease; and if the valley is not full yet, then fear that it will be dry. If Marquises and Kings do not use superior nobility, then fear that they will be overthrown. Therefore, the noble must use the ignoble as their foundation. The high must use the low as their foundation. Therefore, Marquises and Kings call themselves unkindly *'orphans and widowers'*. They use that ignobility incorrectly as their foundation. Therefore, they give too much honour without honour. Hence, do not desire the great splendour of jade, but the grace of natural rock.

§ 40

反也者道之动也.　弱也者道之用也.　天下之物生于有.　有生于无.

Returning is the movement of the Way. Weakness is used by the Way. The world's things originate from Existence. Existence originates from Non-existence.

§ 41

上士闻道，勤而能行之.　中士闻道，若存若亡.　下士闻道，大笑之.　弗大笑不足，以为道.　是以，建言有之曰："明道如昧，进道若退，夷道如类".　上德如浴.　大白如辱.　广德如不足.　建德如偷.　质真若渝.　大方无隅.　大器晚成.　大音希声.　大象无形.　道隐无名.　夫唯道善始且善成.

If competent scholars hear about the Way, then they are able to practise it constantly. If mediocre scholars hear about the Way, then they put it in a safe place and seem to lose it. If incompetent scholars hear about the Way, then they laugh loudly about it. If they did not laugh loudly about it, then they could practice the Way. Therefore, an established saying states: *"The bright Way seems to be obscure, the Way forward seems to be backwards, and the smooth Way seems to be rough"*. Superior virtue is just like a valley. Great pureness seems to be disgrace. Extensive virtue seems to be insufficient. Established virtue seems to drift along. Plain truth seems to change; the greatest square is without edges; the greatest talent matures late; the greatest sound is a rare tone; and the greatest form is without shape. The hidden Way is nameless. Yet, only the Way is good at the beginning and good at the end.

§ 42

道生一. 一生二. 二生三. 三生万物. 万物负阴而抱阳. 冲气以为和. 人之所恶唯孤寡不谷. 而，王侯以自名也. 物或损之而益或益之而损. 故，人之所教议而教人强梁者不得其死. 吾将以为学父.

The Way generated the One. The One generated the Two. The Two generated the Three. The Three generated All-things. All-things carry Yin on their back and carry Yang in their arms. Their balance creates vital energy that restores harmony. People dislike being unrelated orphans and widowers. Yet, Kings and Marquises use those names for themselves. Things may be harmed by benefit and may benefit by being harmed. Therefore, the teachers of humanity discus and teach people that violent people will achieve nothing but death. I will use that as the father of my teachings.

§ 43

天下之至柔驰骋天下之致坚. 无有入于无间. 吾，是以知无为之有益也. 不言之教无为之益天下. 希能及之矣.

The softest of the world will overcome the hardest of the world. What is without substance will penetrate what is without gaps. Therefore, I know that there is benefit in Non-action. Teaching without speaking and Non-action will benefit the whole world. Only a few competent people can attain this.

§ 44

名与身？　孰亲？　身与货？　孰多？　得与亡？　孰病？　什爱必大费.　厚藏必多亡.　故知足不辱.　知止不殆.　可以长久.

Fame or life? What is closer? Life or wealth? What is worth more? Gain or loss? What hurts more? Most people love to spend a lot. The larger their hoard the more they have to lose. Therefore, know what is enough and there will be no disgrace. Know when to stop and there will be no danger. Accordingly, one will endure long.

§ 45

大成若缺.　其用不币. 大盈若冲.　其用不穷. 大直如屈.　大巧如拙.　大赢如绌. 趮胜寒.　静胜热. 清静可以.　为天下正.

The greatest achievement seems to be incomplete. Yet, its usefulness is not reduced. The greatest fullness seems to be empty. Yet, its usefulness is never exhausted. The greatest straightness seems to bend. The greatest skill seems clumsy. The greatest triumph seems insufficient. Activity overcomes the cold. Tranquillity overcomes the heat. Hence, pure tranquillity can be used to regulate the world.

§ 46

天下有道，却走马以粪. 天下无道，戎马生于郊. 罪莫大于什欲. 祸莫大于不知足. 咎莫憯于欲得. 故，知足之为足恒足矣.

If the world possesses the Way, then race horses will be kept for their manure. If the world is without the Way, then war horses will be bred in the suburbs. No greater suffering than having extreme desires. No greater misfortune than not knowing what is enough. No misfortune is more disastrous than the desire to accumulate. Therefore, know that enough is enough and there will be always enough.

§ 47

不出于户以知天下. 不窥于牖以知天道. 其出也弥远者其知弥少. 是以，圣人不行而知. 不见而名. 弗为而成.

Do not leave home in order to learn about the world. Do not look through the window in order to learn about the Way of nature. The more that people travel far away the less they know. Therefore, sages do not travel and yet they know. They do not see and yet they name. They do not act and yet they achieve.

§ 48

为学者日益.　为道者日损. 损之又损以至于无为. 无为而无不为. 取天下也，恒以无事.　及其有事也，不足以取天下矣.

Those who daily pursue knowledge will expand. Those who daily pursue the Way will contract. Contract and contract until there is Non-action left. There is Non-action and yet there is action. If one wants to take the world, then one should always use no effort. When effort is needed, then there is never enough to take the world.

§ 49

圣人无恒心.　以百姓之心为心.　善者善之.　不善者亦善之.　得善也.　信者信之.　不信者亦信之.　得信也.　圣人之在天下也.　歙歙，焉为天下浑心.　百姓皆属其耳目焉.　圣人皆咳之.

Sages are always without opinions. They use the opinions of common people as their opinions. They are good to those who are good. They are also good to those who are bad. So they gain goodness. They trust those who trust them. They also trust those who do not trust them. So they gain trustworthiness. Sages depend on the world. Careful, so that they become merged with the opinion of the world. All common people focus their ears and eyes. Instead, all sages are like children.

§ 50

出生入死.　生之徒十有三.　死之徒十有三.　而民之生生动皆之死地之十有三.　夫何故也?　以其生生也.
盖闻, 善摄生者.　陵行不遇兕虎.　入军不被甲兵.　兕无所揣其角.　虎无所措其爪.　兵无所容其刃.　夫何故也?　以其无死地焉.

Emerging into life is entering into death. Three in ten are companions of life. Three in ten are companions of death. Three in ten people live extremely and move into the realm of death. What is the reason? That is because they live extremely. They are incompetent in hiding, listening, and conserving their lives. Do not walk through the hills in order to meet rhinoceroses and tigers. Do not join the army in order to carry weapons. Rhinoceroses have no place to ram their horns. Tigers have no place to strike their claws. Soldiers have no place to thrust their swords. What is the reason? Because sages avoid the realm of death.

§ 51

道生之而德畜之.　物刑之而器成之.　是以，万物尊道而贵德.　道之尊也德之贵也夫莫之爵也，而恒自然也.
道生之畜之.　长之遂之.　亭之毒之.　养之复之.　生，而弗有也.　为，而弗恃也.　长，而弗宰也.　此之谓玄德.

The Way generates them and virtue raises them. The environment shapes them and competence completes them.

Therefore, All-things respect the Way and admire virtue. Respecting the Way and admiring virtue is not done to obtain a noble position, but it is always done to be natural. The Way generates them and raises them. It grows them and satisfies them. It straightens them and matures them. It supports them and repairs them. Generate but do not possess. Act but do not rely on it. Grow but do not exploit. This is called profound virtue.

§ 52

天下有始以为天下毋. 既得其毋以知其子. 既知其子，复守其毋，没身不殆. 塞其兑闭其门; 终身不勤. 启其兑济其事；终身不救，见小曰明. 守柔曰强. 用其光复归其明. 毋遗身央是谓袭常.

The beginning of the world is the Mother of the world. Obtain the Mother in order to know her children. When knowing her children, return to their nursing Mother, and life will not be in danger. Block the exchange and close the doors; and to the end of life there will be no hard work. Open the exchange and meddle in affairs; and to the end of life there will be no safety. To perceive the small is called brilliant. Following the soft is called strength. Use its light to join its brilliance again. Not losing life to disaster is called following the constant.

§ 53

使我洁有知也． 行于大道，唯施是畏． 大道什夷． 而，民什好径． 朝什除． 田什芜． 仓什虚． 服文彩，带利剑，厌饮食，而货财有余，是谓 ‘盗夸’． 盗夸非道也．

Let me have pure knowledge. When walking on the Great Road, the only thing I fear is action. The Great Road is very smooth. Yet, people prefer the narrow winding roads. Their palaces are very clean. Their fields are overgrown with weeds. Their storehouses are very empty. Those who wear embroidered coloured silk, carry sharp swords, gorge on food, and have a surplus of goods and resources, are called *'boasting thieves'*. Boasting thieves are not the Way.

§ 54

善建者不拔． 善抱者不脱． 子孙以祭祀不辍． 修之于身其德乃真． 修之于家其德有余． 修之于乡其德乃长． 修之于国其德乃丰． 修之于天下其德乃博． 以身观身． 以家观家． 以乡观乡． 以邦观邦． 以天下观天下． 吾何以知天下之然哉？ 以此．

Those who establish it competently cannot be pulled away. Those who embrace it competently cannot be separated. Accordingly, descendants will pay homage forever. Cultivate it in yourself and it will be genuine. Cultivate it in your household and it will be plenty. Cultivate it in your village and its virtue will last long. Cultivate it in your country and its virtue will be abundant. Cultivate it in the world and its virtue will be extensive. Use yourself to examine yourself. Use households to examine households. Use villages to examine villages. Use countries to examine countries. Use the world to examine the world. How do I know that the world is like this? From this account.

§ 55

含德之厚者比于赤子也. 虿，虺，虫地弗螫. 攫鸟弗搏. 骨弱筋柔. 而，握固. 未知牝牡之合. 而，朘怒. 精之至也. 终日号. 而，不嚘. 和之至也. 知和曰常. 知常曰明. 益生曰祥. 心使气曰强. 物壮则老. 谓之不道. 不道早已.

Those who have substantial virtue could be compared to new-born babies. Scorpions, vipers, and insects do not bite them. Birds of prey will not seize them. Their bones are weak and their tendons soft. Yet, their grip is firm. They do not know about the joining of male and female. Yet, their male organ is vigorous. Their energy is optimal. They cry all day. Yet, they do not become hoarse. Their harmony is optimal. Knowing harmony is called the constant. Knowing the constant is called brilliant. Benefiting life is called fortune. Using the mind's vital energy is called powerful. Strong things will become weak. They are not called the Way. What is not the Way will perish soon.

§ 56

知者弗言也． 言者弗知也． 塞其兑闭其门． 挫其锐． 解其纷，和其光，同其尘． 是谓玄同． 故不可得而亲也． 亦不可得而疏． 不可得而利． 亦不可得而害． 不可得而贵． 亦，不可得而贱． 故，为天下贵．

Those who know do not speak. Those who speak do not know. Block the exchange and close the doors. File their sharpness. Untangle their disorder. Soften their glare. Merge their dust. This is called profound unification. It cannot be achieved by attachment. Neither can it be achieved by detachment. It cannot be achieved by benefit. Neither can it be achieved by harm. It cannot be achieved by admiration. Also, it cannot be achieved by contempt. Therefore, the world admires it.

§ 57

以正治国． 以奇用兵． 以无事取天下． 吾何以知天下其然也？ 才． 夫天下多忌讳，而民弥贫． 民多利器，而邦家滋昏． 人多知，而奇物滋起． 法物滋章而盗多有． 是以，圣人之言曰：“我无为也而民自化． 我好静而民自正． 我无事而民自富． 我欲无欲而民自朴”．

Use justice when ruling the State. Use surprise when employing armies. Use no effort when taking the world. How do I know it is like this? Only after this. The more prohibitions there are in the world, the poorer the people will be. The more sharp weapons people have, the more the

State's household will be confused. The more people know, the stranger the things they begin to develop. Rules increase the content of the law and, therefore, there will be more criminals. Hence, the saying of the sages states: *"I practice Non-action and the people will transform themselves. I am tranquil and the people will perfect themselves. I use no effort and the people will become wealthy by themselves. I desire not to desire and the people will become simple by themselves"*.

§ 58

其正闵闵，其民屯屯. 其正察察，其民缺缺祸. 祸福之所伏. 福祸之所伏. 孰知其极？ 其无正也. 正复为奇. 善复为妖. 人之迷也其日固久矣. 是以，圣人妨而不割，廉而不刺，直而不绁，光而不耀.

If the laws are very lax, then the people will have extreme surpluses. If the laws are very strict, then the people will have extreme shortages and misfortune. Misfortune is fortune's place to hide. Fortune is misfortune's place to hide. Who knows their extremes? There is no normal. Normal turns around and becomes abnormal. Good turns around and becomes evil. That has confused everybody for a long time. Therefore, sages are interfering but not cutting, sharp but not stabbing, straight but not rigid, and bright but not dazzling.

§ 59

治人事天莫若啬. 夫唯啬是谓早服. 早服是谓重积德. 重积德，则无不克. 无不克，则莫知其极. 莫知其极，可以有国.　有国之母可以长久.　是谓, 深根固氐. 长生久视之道也.

In ruling people and working with nature there is nothing like frugality. Only thosc who are frugal are called to early service. Prepare for the call to serve early by a significant accumulation of virtue. If there is a significant accumulation of virtue, then nothing is impossible. If nothing is impossible, then one's limits are unknown. If one's limits are unknown, then one may possess the country. Possess the Mother of the country and accordingly endure long. That is called: having deep roots and a strong foundation. Having a long life through a lasting regard for the Way.

§ 60

治大国若亨小鲜. 以道莅天下，其鬼不神.　非其鬼不神也，其神不伤人也.　非其神不伤人也.　圣人亦弗伤人也.　夫两不相伤. 故，德交归焉.

Ruling a large country is like enjoying small delicacies. Use the Way to attend to the world, then spirits will have no power. It is not that spirits have no power, but their power will not harm people. Not that power cannot harm people. Sages do not harm people either. Both do not harm each other. Therefore, virtue unites and returns.

§ 61

大国者天下之下流也.　天下之交也.　天下之牝也. 牝恒以静胜牡.　以其静也.　故宜为下也.　故，大国以下小国，则取小国．　小国以下大国，则取于大国．故，或下以取．　或下而取．　故大国者不过欲兼畜人．小国者不过欲入事人．　夫皆得其欲．　故，大者宜为下．

A large country should take a low position. It is the intersection of the world. It is the female of the world. The female always uses tranquillity to overcome the male. She is tranquil. Therefore, she is better in a low position. Hence, a large country should use a lower position than a small country, when associating with that smaller country. A small country should use a lower position than a large country, when associating with that larger country. Therefore, some might take the lower position to associate. Others might be in the lower position to associate. Those in a large country only desire to merge and raise people. Those in a small country only desire to join other business people. They all obtain what they desire. Therefore, the larger one better acts as the lowest one.

§ 62

道者万物之注也．　善人之保也．　不善人之所保也．美言可以市．　尊行可以贺人．　人之不善也何弃之有．故，立天子置三卿，虽有共之璧以先四马．　不善坐而进此．　古之所以贵此道者何也？　不谓：“求以得有罪以免与”．　故，为天下贵．

People and All-things concentrate on the Way. It is the protection for competent people. It is the sanctuary for incompetent people. Pleasing words might be exchanged. Respectful conduct might honour people. Incompetent people forsake their existence unnecessarily. Therefore, when the Emperor is crowned and the three ministers are installed, a mill-stone of jade preceded by four horses, is not as good as sitting down and presenting this. So, why did the Ancients value the protection of this Way? Did they not say: *"Use it to obtain what you seek and use it to escape what you suffer"*. Therefore, it is valuable for the whole world.

§ 63

为无为. 事无事, 味无味. 大小多少. 报怨以德. 图难, 乎于其易. 为大, 乎于其细也. 天下之难作于易. 天下之大作于细. 是以, 圣人终不为大. 故, 能成其大. 夫轻诺必寡信. 多易必多难. 是以, 圣人猷难之. 故, 终于无难矣.

Act with Non-action. Work without effort. Taste without savouring. Make the large small and the many few. Repay hatred with kindness. Pursue the difficult, while it is easy. Act large, while it is still small. The world's most difficult things arise from the most easy ones. The world's largest things arise from the smallest ones. Therefore, all sages will avoid great actions. Hence, they can achieve greatness. Those who make rash promises are certainly difficult to trust. Those who regard everything as easy will have certainly many difficulties. Therefore, sages regard everything as difficult. Hence, they have no difficulties in the end.

§ 64

其安也易持也. 其未兆也易谋也. 其脆也易破也. 其微也易散也. 为之于其未有也. 治之于其未乱也. 合抱之木作于毫末. 九成之台作于蔂土. 百仞之高始于足下. 为之者败之. 执之者失之. 是以, 圣人无为也. 故, 无败也. 无执也. 故, 无失也. 民之从事也, 恒于其几成事也而败之. 故, 慎终若始. 则无败事矣. 是以, 圣人欲不欲而不贵难得之货. 学不学而复众人之所过. 圣人能辅万物之自然, 而弗敢为.

That what is at rest is easy to hold. That what is not manifest is easy to plan. That what is fragile is easy to break. That what is small is easy to scatter. Act when it has not happened yet. Control it when it is not chaotic yet. A tree that takes both arms to embrace grows from a little cutting. Nine-tenth of a tower rises from a simple basket of earth. A thousand meters height starts from under your feet. Those who act will fail. Those who hold will lose. Therefore, sages will use Non-action. Therefore, they will not fail. They will not hold. Therefore, they will not lose. In handling their affairs, people fail often close to their success. Therefore, be as careful at the end as at the beginning. Then affairs will not fail. Therefore, sages desire not to desire and do not admire goods that are difficult to obtain. They learn not to learn and repair the mistakes of others. Sages complement the nature of All-things, but they do not dare to act.

§ 65

古之故曰：“为道者非以明民也”． 将以愚之也． 民之难治也以其知也． 故，以知治邦邦之贼也． 以不知治邦邦之德也． 恒知此两者． 亦稽式也． 恒知稽式是谓玄德． 玄德深矣． 远矣与物反矣． 乃至大顺．

Therefore, the Ancients said: *"Do not use the Way to enlighten people"*. Instead, use it to keep them simple. People are difficult to rule if they use their knowledge. Therefore, using knowledge to rule the country is betraying the country. Using no knowledge to rule the country is benefiting the country. Remember always those two things. Examine also the principle. Always remembering to examine the principle is called profound virtue. Profound virtue is deep. Even far away things return to it. The great order is perfect.

§ 66

江海之所以能为百浴王者？ 以其善下之也． 是以，能为百浴王． 是以，圣人之欲上民也必以身下之． 其欲先民也必以其身后之． 故，其居上而民弗重也． 其居前而民弗害也． 天下皆乐推而弗厌也． 非以，其无争也． 故，天下莫能与之争．

How are the river and the sea able to be the Kings of a hundred valleys? They use competently their low position. Therefore, they are able to be Kings of a hundred valleys. Therefore, sages who desire to be above the people must place themselves below them. Those who desire to lead people must place themselves behind them. Therefore, they stay above and the people will not weight them down. They stay in front and the people will not harm them. Everyone in the world will be happy to elect them without objection. Having no purpose, they do not strive. Therefore, the world cannot strive with them.

§ 67

天下皆谓我大. 大而不肖. 夫唯不肖故能大. 若肖矣，其细矣. 我恒有三宝持而保之. 一曰慈. 二曰俭. 三曰不敢为天下先. 夫慈故能勇. 俭故能广. 不敢为天下先故能为成事长. 今，舍其慈且勇；舍其俭且广；舍其后且先；则必死矣. 夫慈以战则胜. 以守则固. 天将救之如以慈垣之.

Everyone in the world calls me great. Great and different. Only those who are different can be great. If they were similar, then they would be insignificant. I have always three treasures that I keep and protect. The first one is called compassion. The second one is called frugality. The third one is called humbleness. Those who are compassionate can be courageous. Those who are frugal can be generous. Those who do not dare to act as the first of the world can be successful leaders of affairs. Now, those who abandon compassion and are yet courageous; abandon their frugality and are yet generous; abandon their humbleness and are yet leading; they will certainly die. Those who use compassion to attack will triumph. Those who use it to defend will stand firm. Nature will protect them with a wall of compassion.

§ 68

善为士者不武也. 善单者不怒. 善胜敌者弗与. 善用人者为下. 是谓不争之德. 是谓用人. 是谓配天. 古之极也.

Competent warriors do not like war. Competent chiefs will not get angry. Competent conquerors will not engage. Competent leaders will take a low position. That is called the virtue of not striving. That is called employing people. That is called matching with nature. It is the ultimate principle of the Ancients.

§ 69

用兵有言曰: "吾不敢为主, 而为客. 吾不敢进寸, 而退尺". 是谓: 行无行. 攘无臂. 乃无敌矣. 执无兵. 祸莫大于无敌. 无敌近丧吾宝矣. 故, 抗兵相若则哀者胜矣.

Warriors have a saying that states: *"I do not dare to act as a host, but act as a guest. I do not dare to advance an inch, but retreat a foot"*. That is called: moving without moving. Rolling up the sleeves without showing an arm. Be without resistance. Hold without weapons. No greater misfortune than meeting no resistance. Meeting no resistance is close to losing my treasures. Therefore, when equal armies face each other the reluctant one will win.

§ 70

吾言什易知也什易行也． 而，人莫之能知也而莫之能行也． 夫言有宗事有君，夫唯无知也． 是以，不我知． 知我者希则我贵矣． 是以，圣人被褐而褱玉．

My words are very easy to understand and very easy to apply. Yet, people cannot understand them and they cannot apply them. My words have precedence and my affairs have a sovereign, but they do not understand that. Therefore, they do not understand me. I am valuable to those few who understand me. Therefore, sages wear cheap cloth and conceal jade.

§ 71

知不知尚矣． 不知不知病矣． 是以，圣人之不病也．以其病病也． 是以，不病．

Knowing that you do not know earns respect. Not knowing that you do not know is a weakness. Therefore, sages are not weak. They consider their weakness as a weakness. Hence, it is not a weakness.

§ 72

民之不畏威，则大威将至矣． 毋狎其所居． 毋厌其所生． 夫唯弗厌，是以不厌． 是以，圣人自知，而不自见也． 自爱，而不自贵也． 故，去彼而取此．

When people do not fear authority, then greater authority will appear. Do not take their dwellings by force. Do not reject them a place to live. Only, if they are not rejected, then they will not reject you. Therefore, sages know themselves, but do not display themselves. They love themselves, but do not admire themselves. Therefore, reject that and accept this.

§ 73

勇于敢者则杀. 勇于不敢者则活. 此两者或利或害. 天之所亚. 孰知其故? 天之道不争, 而善胜. 不言, 而善应. 不召, 而自来. 单而善谋. 天网恢恢. 疏而不失.

Those who are courageous in daring will be killed. Those who are courageous in not daring will live. Those two things could be beneficial or could be harmful. Nature takes a low place. Who knows its reason? The Way of nature is not to strive, but to overcome through competence. Without speaking, it answers with competence. Without calling, things come. It is simple and plans with competence. The net of nature is very extensive. It dredges and nothing escapes.

§ 74

若民恒且不畏死, 若何以杀惧之也? 若民恒畏死而为畸者, 吾将得执而杀之夫孰敢矣? 若民恒且必畏死, 则恒有司杀者. 夫代司杀者, 杀是代大匠斲也. 夫代大匠斲者则希有不伤其手矣.

If people never fear death, how could the use of executions scare them? If people always fear death and act abnormal, how could I dare to seize, hold, and execute them? If people always fear certain death, then there is always someone in charge of executing them. Those who act on behalf of the one that is in charge of the executions, execute as if they act on behalf of the Master Carpenter. From those who act on behalf of the Master Carpenter only a few will not cut their hands.

§ 75

人之饥也，以其上食税之多也． 是以，饥． 百姓之不治者也，以其上之有以为也． 是以，不治． 民之轻死也，以其求生之厚也． 是以，轻死． 夫唯无以生为者是贤于贵生也．

People are hungry, because their food taxes are high. Therefore, they are hungry. Common people cannot be governed, because their leaders act for their own purposes. Therefore, the people cannot be governed. People take death lightly, because they seek to live substantially. Therefore, they take death lightly. Only those who do not act for the purpose of living are knowledgeable at valuing life.

§ 76

人之生也柔弱． 其死，也坚强． 万物，草木之生也柔脆． 其死，也枯槁． 故，曰坚强者死之徒也． 柔弱生之徒也． 是以，兵强则不胜． 木强则折． 故，强大居下． 柔弱居上．

People are born soft and weak. They die, hard and strong. All-things, grasses, and trees are born soft and fragile. They die, dry and brittle. Therefore, the hard and strong are called companions of death. The soft and weak are the companions of life. Hence, a strong army will not win. A strong tree will be broken. Therefore, the strong and big occupy the low positions. The soft and weak occupy the high positions.

§ 77

天之道其犹张弓者也. 高者抑之. 下者举之. 有余者损之. 不足者与之. 故，天之道损有余而补不足. 人之道则不然. 损不足而奉有余. 夫孰能有余奉不足于天下者乎唯有道者乎. 是以，圣人为, 而弗恃. 成功，而弗居也. 其不欲见贤也.

The Way of nature is like flexing a bow. High things are lowered. Low things are raised. It takes from those who have plenty. It gives to those who have not enough. Therefore, it is the Way of Nature to take from what is plenty and give to what is not enough. However, the Way of People is different. They take from what is not enough and give to what is plenty. Hence, those who have plenty and give to those in the world who have not enough are the only ones who possess the Way. Therefore, sages act, but do not rely on it. They succeed, but do not claim. They do not desire to display their knowledge.

§ 78

天下莫柔弱于水. 而，攻坚强者，莫之能胜也. 以其无以易之也. 故，柔之胜刚也. 弱之胜强也. 天下莫弗知也.　而，莫能行也.　故，圣人之言云：“受邦之垢是谓社稷之主.　受邦之不祥是谓天下之王”.　正言若反.

Nothing in the world is as soft and weak as water. Yet, in attacking hard and strong things, nothing has a greater ability to overcome them. Therefore, nothing could replace its purpose. The softest will overcome the hardest. The weakest will overcome the strongest. Nobody in the world does not know this. Yet, nobody does practise it. Therefore, sages have a saying that states: *"Accept the country's shame and be called the Kingdom's leader. Accept the country's misfortune and be called the world's King"*. True words seem to be paradoxical.

§ 79

和大怨必有余怨.　焉可以为善？　是以，圣人执右契而不以责于人.　故，有德司契.　无德司彻.　夫天道无亲.　恒与善人.

Calm a great hate and certainly some hate will remain. How could this be considered competent? Therefore, sages adhere to orthodox agreements and do not obligate other people by purpose. Therefore, those with virtue will uphold the agreement. Those without virtue will uphold the details. The Way of Nature has no favourites. It is always with the competent people.

§ 80

小邦寡民，使有十百人之器，而毋用也．　使民重死而不远徙．　有车无所乘之．　有甲兵无所陈之．　使人复结绳而用之．　甘其食美其服．　乐其俗安其居．　邻邦相望．　鸡狗之声相闻．　民至老死不相往来．

In a small country with a few people, let everybody have many tools, but not use them. Let the people be serious about death and not move far away. Let them have carriages without using them. Let them have weapons without displaying them. Let them return to knotted cords and use them. Sweeten their food and beautify their clothes. Enjoy their customs and secure their dwellings. Neighbouring countries will see each other in the distance. They hear each other's chickens and dogs. Nevertheless, the people die of old age and have never visited each other.

§ 81

信言不美．　美言不信．　知者不博．　博者不知．　善者不多．　多者不善．　圣人无积．　既以为人，己俞有．既以予人矣．　己俞多．　故，天之道利而不害．　人之道为而弗争．

True words are not pleasing. Pleasing words are not true. Those who know are not educated. Those who are educated do not know. Those who are competent have not much. Those who have much are not competent. Sages do not hoard. Since they are used to act for other people, they receive more possessions for themselves. Since they are used to give to other people, they have much more themselves. Therefore, the Way of Nature is beneficial and without harm. Accordingly, the Way of People should be action without strive.

This version
of Lao Zi's Dao De Jing
is based on [7] [13] [14] [15] [16].

6.2 Ancient Theory Challenges Modern Physics

Dr Auke J. Schade

Unpublished manuscript 9 February 2021

Abstract

Two-and-halve thousand years ago, Lao Zi wrote Dao De Jing. His demystified book shows that he was the first to describe scientific concepts including: gravity, antimatter, nothingness, fluctuation, and a zero-sum creation theory. His creation Theory of Fluctuating Nothingness holds that Nothingness was divided into the mirror universes of Matter and Antimatter. Their fluctuation around Nothingness creates a temporal buffer that prevents their mutual annihilation and conceals Antimatter from perception. The constancy of Nothingness underlies the cosmic balance between matter and antimatter. That balance is maintained by the force Qi, which comprises the gravitational, electro-magnetic, and nuclear forces. Movements such as reactions, entropy, and chaos are manifestations of that balancing process. Lao Zi's ancient theory challenges the history, purpose, methodology, and philosophical premises of modern physics. The purpose of Dao or physics is obtaining knowledge about nature. On the other hand, the purpose of De or psychology is obtaining wisdom in order to apply that knowledge productively. Lao Zi's theories are relative to our era, because they foster crucial research in several scientific fields that could help the next generation to face manmade extinction problems.

Lao Zi's Dao De Jing

The Chinese philosopher Lao Zi (~570-490 BCE) wrote the book Dao De Jing, which means: A Classic about the Physics of Psychology [1] [2] [3] [4] [5].[1] However, his theories about Dao or physics, and De or psychology, were developed by a civilization he called: The Ancients ([2] §15 & 22); ([3] §62, 65 & 68).[2] Lao Zi's warns about the sophistication of those theories: *"My words are very easy to understand and very easy to apply. Yet, people cannot understand them and they cannot apply them"* [1, p. §70].[3]

Zhuang Zi (~369-286 BCE) and subsequent scholars mystified unintentionally Dao De Jing with their attempts to explain Lao Zi's theory of physics in super-natural terminology [6] [7]. Physics needed to recover, before the knowledge of the Ancients could be recognized. In accord, modern physics is the scientific basis for the presented demystification of Lao Zi's Dao De Jing [4] [5].

A translation-interpretation process of several distinct stages was applied to demystify Lao Zi's Dao De Jing. (I) A special dictionary was compiled containing the exhaustive English meanings for each pictograph mentioned in six reputable Chinese versions of Dao De Jing [1].[4] (II) The translations from Lao Zi's archaic Chinese pictographs to modern ones were validated by integrating those six Chinese versions [2] [3]. (III) The translation from modern Chinese to English was supported by comparing the integrated Chinese text with nine reputable English translations. To determine the final meaning of each pictograph, the consistency of that meaning was maximized across the entire Dao De Jing [2] [3]. (IV) To clarify Lao Zi's theories, the holistic format of Dao De Jing was reorganized into a rational format organized by topic [4]. (V) Lao Zi's ancient jargon was updated to the jargon of modern physics [4].[6] (VI) Lao Zi's theory of physics was reconstructed and named: Theory of Fluctuating Nothingness [5]. (VII) Lao Zi's theory of psychology was reconstructed and named: Cognitive Criteria Theory [5].

Lao Zi explains that the main characteristics of Dao are: Infiniteness, Oneness, and Nothingness.[7] He wrote about the Infiniteness of Dao: *"Above the One (Dao) there is no void. Below it (Dao) there is no substance. It (Dao) is infinite".* ([8] §14).[8] Therefore, there is nothing else than Dao. Furthermore, Lao Zi explains the Oneness of Dao. If anything is divided into smaller parts, then we need names to identify those parts. Hence, naming is a synonym for dividing, while nameless is a synonym for Oneness. In accord, Lao Zi wrote: *"It (Dao) cannot be named ([8] §14). The Way (Dao) is forever Nameless* ([8] §32 & 37). *The hidden Way (Dao) is Nameless"* ([9] §41).[9] In addition, Lao Zi describes the Nothingness of Dao: *"Every time it (Dao) returns to Nothingness. It (Dao) is called shapeless. Like the shape of Nothingness* ([8] §14).[10] *There was a thing (Dao) Undivided and complete before the Sky and the Earth were born. Desolate. Empty (Nothingness)"* ([8] §25).[11] Lao Zi's Nothingness cannot change, because there is nothing else to change it with. Therefore, Dao is the cosmic infinite oneness of constant Nothingness.

Lao Zi's Theory of Physics

Lao Zi proposes that Dao is the principle, substance, and force that explains the origin, creation, and working of the universe. He wrote about the origin of the universe: *"There was a thing Undivided (Oneness) and complete (Infiniteness) before the*

Sky and the Earth were born. Desolate. Empty (Nothingness). Independent and unchanging (Constant)... Yet, it acts as the origin of the world. I do not know how its name is pronounced, but I call it the Way (Dao)" ([8] §25).[12] Lao Zi does not pretend to know the origin of Dao: *"I do not know whose child it (Dao) is, but it seems to predate the Emperor (Humanity)".* ([8] §4).[13] *In order to study the origin: "Use it (Dao) to understand its (Dao's) ancient origin".* ([8] §14).[14] Similarly, modern scientists study the present universe in order to learn about its origin.

Lao Zi wrote about the creation of the universe: *"The One (Nothingness) generated the Two".* ([9] §42).[15] In addition, he explained: *"The world's things originate from Existence. Existence originates from Non-existence".* ([9] §40).[16] Existence and Non-existence could be referred to as Matter and Antimatter [10]. Therefore, substitution provides: Nothingness generated Matter and Antimatter. The cosmos contains the entire amount of Matter and Antimatter, while our universe is that part of the cosmos that contains only Matter. The cosmic constancy of Nothingness forces Matter and Antimatter to maintain a balance of equal opposites. Hence, the sum of Matter (+M) and Antimatter (–M) equals Nothingness (0). Therefore: $(+M) + (-M) = 0$.

Figure 1: Lao Zi's *Zero-sum Creation Theory*

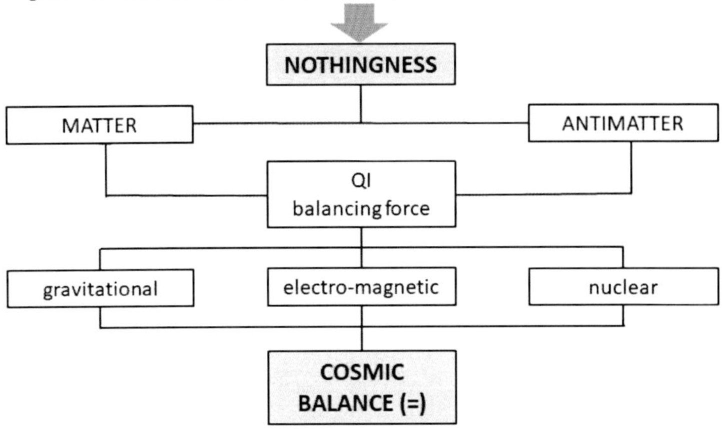

Figure 1 shows Lao Zi's *'Zero-sum Creation Theory'*, which holds that the constant cosmic Nothingness was divided in equal opposites of Matter and Antimatter. Furthermore, Qi is the manifestation of Dao as the cosmic force that maintains the balance. That generic force manifests itself specifically as gravitational, electro-magnetic, and nuclear forces. Those specific forces maintain the cosmic balance, while Qi is the force that unifies them. Furthermore, Lao Zi's cosmic principle of balance is supported by the equal sign (=), which is fundamental to all modern sciences, arithmetic, mathematics, accountancy, and logic. Consequently, even if his Zero-sum Creation Theory could be rejected, then his principle of cosmic balance would be still valid.

Lao Zi's Zero-sum Creation Theory has two major problems. Firstly, his theory predicts mirror universes of Matter and Antimatter of equal magnitudes. However, the location of the Antimatter universe has not been discovered yet. Secondly, contact between Matter and Antimatter would lead to their mutual annihilation. Therefore, some physicists hypothesized that a Leidenfrost-layer separates those universes [10].

Lao Zi solves both problems eloquently with his concept of *'Returning'*.[17] Figure 2 illustrates that Lao Zi's concept refers to the modern notions of fluctuation and wavelength. He explains: *"Every time it (Dao) returns to Nothingness* ([8] §14).[18] *All-things around us rise and I watch them return. Those things are numerous and each one returns to its roots. Returning to the roots is called tranquillity (balance). Tranquillity is called returning to order. Returning to order is a constant* ([8] §16).[19] *Continuous means forever. Forever means returning"* ([8] §25).[20] His philosophy implies that each particular state of existence returns to the previous one. For example, each day returns to night, while each summer returns to winter and vice versa. Therefore, both the magnitudes of light and temperature show waveforms if plotted across time. That suggests that Fluctuating would be the modern term for Lao Zi's ancient concept of Returning. Lao Zi concludes: *"Returning is the movement of the Way"* ([9] §40).[21] The Way is a synonym for Dao and its manifestation of Nothingness. Substitution provides: "*Fluctuation is the movement of Nothingness.*"

As shown in Figure 2, fluctuation turns the static balance of Matter and Antimatter into a dynamic one that alternates them across time. Hence, Lao Zi's *'Fluctuating Nothingness Theory'* holds that the cosmos consists of mirror universes of Matter and Antimatter fluctuating in the same spatial location around the infinite oneness of constant Nothingness.[22]

Figure 2: Fluctuation of Matter and Antimatter around Nothingness

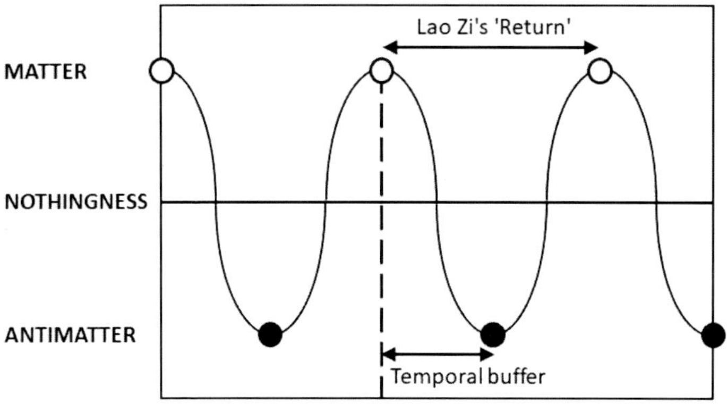

Lao Zi's idea of fluctuation has significant implications. The Temporal Buffer between the mirror universes of Matter and Antimatter protects those universes from mutual annihilation. Furthermore, Matter and Antimatter are relative concepts determined by the state of fluctuation during the moment of observation. The observer would fluctuate in synchrony with the cosmos between the states of Matter and Antimatter. When the cosmic fluctuation would change Matter into Antimatter, the observer would perceive the former Antimatter as Matter, because all references change simultaneously. The reverse occurs when Antimatter is changed to Matter. Without awareness, the observer fluctuates between the universes of Matter and Antimatter. Hence, the observer will never perceive the universe of Antimatter. What is observed as Antimatter, within the apparent universe of Matter, would be out-of-sync fluctuation. The frequency of the fluctuation drives the cosmic clock that creates the arrow of time.

The cosmic balance between Matter and Antimatter is reflected in the multitude of interconnected Yin-Yang balances that constitute our universe. Yin means literally: shady side of the hill. However, in Lao Zi's philosophy, Yin is a generic term that represents a cluster of negative concepts. On the other hand, Yang means literally: *sunny side of the hill.*[23] Yang is a generic term that represents a cluster of positive concepts. As a result, Yin-Yang balances include contrasts such as: black-white; female-male; weak-strong; below-above; emotion-logic; retreat-advance; dispose-accumulate; passive-active; conceal-reveal; accept-reject; etc. [11]. Lao Zi wrote: *"All-things carry Yin on their back and carry Yang in their arms"* ([9] §42).[24] That means that the Yin-Yang balances permeate everything from subatomic particles to mountain ranges, and from inorganic to organic material.

Lao Zi continues: *"Any difference creates vital energy (Qi) that restores their harmony (Yin-Yang balance)"* ([9] §42).[25] However, the Yin-Yang balances and the restoring force Qi are normally hidden from our perception. Qi is only detectable when it is restoring balances. For example, we do not notice the resting gravitational balance, but we notice the avalanche restoring the disturbed balance. We do not notice the resting electro-magnetic balance, but we notice the lighting and thunder restoring the disturbed balance. We do not notice the resting atomic balance, but we notice the nuclear explosion restoring the disturbed balance. Disturbing the balance is in principle counterproductive, because it will be undone by the unstoppable force Qi.

Figure 3: Lao Zi's *Theory of Fluctuating Nothingness*			
DAO ➡ ⬇	**Origin Universe**	**Creation Universe**	**Working Universe**
Substance	Nothingness	Matter & Antimatter	Matter
Principle	*(unknown)*	Fluctuating Nothingness	Yin-Yang Balances
Force	*(unknown)*	Qi *(Big Bang)*	Qi gravity *(electro-magnetic, nuclear)*

Figure 3 shows the diagram summarizing Lao Zi's Theory of Fluctuating Nothingness. Despite Lao Zi's sophisticated theory, serious questions about the origin of the cosmos remain. For example: What causes the fluctuation? How could there be fluctuation without a fluctuating medium? If there is a medium, where did it come from? Does the elusive God-particle originate ultimately from Nothingness?

Similarities with Modern Physics

Sir Isaac Newton (1642-1727) described gravity in his Principia [12]. On the other hand, Lao Zi wrote in Dao De Jing: *"The Immortal Valley Spirit is called the Mysterious Female"* ([8] §6).[26] That seems to be esoteric. However, immortal means eternal; valley is a Yin symbol for downwards; while the Chinese character for spirit means also force [11].[27] Hence, Immortal Valley Spirit means simply the Eternal Downward Force. That is Lao Zi's perfect definition of Newton's gravity. Furthermore, gravity itself cannot be seen and therefore it is mysterious, while female is another Yin symbol for downwards. Hence, Lao Zi calls Gravity the Mysterious Female. Substitution provides modern physics: The eternal downward force is called gravity. Although Lao Zi used gravity to explain Qi, he might have known also about the more obscure electro-magnetic and nuclear forces. He wrote: *"Inside (Dao) there are things. Hidden. Obscure. Its (Dao's) centre has energy. Its energy is very real"* ([8] §21).[28]

Newton presented his Second Law of Thermodynamics in his Principia [12]. That law holds that entropy will always increase in a closed system, or alternatively, the order of a closed system will always decrease. In addition, Stephen Hawking mentioned that entropy is the principle of the universe [13]. However, Lao Zi's theory holds that entropy is the observable restoration of hidden Yin-Yang balances by the force Qi: *"Returning to order (balance) is a constant" (§16).* Hence, Lao Zi's balance is the rationale for Newton's entropy.

Newton presented his Third Law of Motion in his Principia [12]. That law predicts that each action evokes a reaction of equal magnitude and opposite direction. Instead, Lao Zi theory holds that reality consist of a multitude of hidden Yin-Yang balances interacting as domino-chains. Each action will be undone by the restoring force Qi. Hence, Lao Zi's balance is the rationale for Newton's law. In addition, Lao Zi's theory implies that the magnitude of the reaction (Qi) depends on the characteristics of the particular Yin-Yang balance, rather than on the magnitude of the action. For example, rolling a tiny snowball down a mountain could create a destructive avalanche. Similarly, disturbing the natural balances with relatively small amounts of manmade

CFC, DDT, and CO2 evoked gigantic reactions. Therefore, Lao Zi warned people that disturbing Yin-Yang balances is counterproductive: *"Concentrate on removing extremes. Nurture tranquillity (balance) diligently"* ([8] §16).[29] Unfortunately, humanity has ignored his warnings. As a result, the next generation is facing manmade extinction problems.

Albert Einstein (1879-1955) published in 1905 four papers that revolutionized Newtonian physics [14] [15] [16] [17]. Similar to Lao Zi, Einstein was a psychophysicist, because his concept of relativity returned the psychology of the observer back into physics. Furthermore, Einstein's famous equation (E = mc2) shows the equivalence (Oneness) of energy (E) and mass (m) [18]. However, Lao Zi's earlier theory holds that the entire cosmos consists of Dao in its manifestation of Oneness. Therefore, Einstein's mass-energy equivalence is a special case of Lao Zi's cosmic Oneness. In addition, the equal sign (=) in Einstein's equation, supports Lao Zi's principle of cosmic balance.

Paul Dirac (1902-1984) predicted in 1928 the existence of positively charged electrons. That resulted in an equation describing both Matter and Antimatter [19] [20]. In 1932, Carl Anderson confirmed Dirac's prediction [21] [22]. However, Lao Zi had much earlier proposed: Nothingness generated Matter and Antimatter (§42).

Erwin Schrödinger (1887-1961) coined the term Quantum-entanglement, which Einstein called the Spooky action at a distance [23] [24]. In 1935, Einstein, Podolsky, and Rosen published a pivotal paper about quantum-entanglement [25]. Experiments have supported the existence of quantum-entanglement of photons, neutrinos, molecules, and diamonds [26] [27] [28] [29] [30]. However, Lao Zi's cosmic Oneness accounts for the oneness observed in Schrödinger's quantum-entanglement.

Edward Lorenz (1917-2008) published in 1963 a seminal paper about Chaos Theory [31]. Contradicting Newton's Third Law of Motion; Lorenz proposed that the flapping wings of a small butterfly could cause a gigantic tornado elsewhere (Butterfly Effect). However, Lao Zi's theory holds that chaos is the observable restoration of hidden Yin-Yang balances by Qi: *"Returning to order (balance) is a constant" (§16).* Therefore, Lao Zi's balance is the rationale for Lorenz' chaos. In addition, the magnitude of the reaction (Qi) depends on the characteristics of the particular Yin-Yang balance, rather than on the magnitude of the action. Hence, Lao Zi's theory accounts for Lorenz' Butterfly Effect.

Fritjof Capra (1939--) mentioned in his book The Tao of Physics that particle physicists have found a point where the breakdown of particles does not result in any smaller pieces. Those particles behave in a dynamic energy pattern where they seem to recreate themselves following collisions with other particles [32]. However, Lao Zi's theory holds that the entire cosmos consists of Dao in its manifestation of Oneness. Hence, Lao Zi's Cosmic Oneness is the rationale for Capra's Subatomic Oneness.

Stephen Hawking (1942-2018) mentioned that concentrating all the positive and negative energy of the universe in one space-point would create zero energy (zero-energy universe) [13]. However, Lao Zi's theory holds that the cosmos was created from Nothingness. Hence, concentrating the entire cosmos in one space-time-point would create a zero-cosmos.

Consequently, Hawking's idea is a special case of Lao Zi's cosmological theory.

Edward Tryon (1940-2019) published in 1973 the paper: Is the Universe a Vacuum Fluctuation? [33]. Tryon's mathematical model explains how fluctuation might have formed the universe from a vacuum (Nothingness) without violating the laws of physics. However, Lao Zi proposed much earlier: *Nothingness generated Matter and Antimatter* ([9] §42). In addition, Lao Zi's idea of Returning is a synonym for Tryon's Fluctuation. Consequently, Lao Zi's Theory of Fluctuating Nothingness precedes Tryon's Vacuum Fluctuation Theory.

Differences with Modern Physics

In Lao Zi's philosophy, physics is the only valid and reliable basis for psychology. His aim of physics is maximizing success: *"Use it (Dao) to obtain what you seek and use it to escape what you suffer"* ([9] §62).[30] Unfortunately, people are the lowest in the natural hierarchy: *"People follow the Earth; the Earth follows the Sky; the Sky follows the Way (Dao); and the Way (Dao) follows Nature (Dao)"* ([8] §25).[31] As a result, people are conditioned by the immutable laws of nature as rats in a Skinner-box: *"What is not the Way (Dao) will perish soon* ([8] §30; [9] §55).[32] *The Way of Nature (Dao) has no favourites. It is always with the competent people"* ([9] §79).[33] Therefore, competent people learn to align with Dao: *"It (Dao) is the protection for competent people"* ([9] §62).[34] Humanity ignored Lao Zi's warnings. It disturbed the natural balance and opposed the unstoppable Qi. As a result, Yin-Yang balances are cascading like chains of falling dominos. Manmade extinction problems, such as nuclear war and climate change, suggest that modern physics has become complicit in some threats to the survival of humanity. Physicists have to ask: Will humanity align with Qi? Are scientists morally responsible for the manmade extinction problems?

Lao Zi was the first scientist to synthesize objective physics and subjective psychology into psychophysics. On the other hand, Newtonian physics attempted to be purely objective. It did not take into account the subjective perceptions and sensations of the observer. For example, Newton's prism experiment shows the limitations of that approach. The prism refracts white light into a colour spectrum [34]. However, in the external physical reality there are only frequencies and wavelengths. The colours perceived by the observer are created within the mind. Therefore, the prism demonstrates predominantly subjective psychological effects, rather than objective physical ones. Albert Einstein extended Newtonian physics with his Relativity Theory [35]. In that theory, the subjective perceptions and sensations of the observer affects the understanding of physics. Therefore, Einstein returned to Lao Zi's psychophysics. Physicists have to ask: Should modern physics be synthesized with psychology? Is the modern definition of psychophysics too narrow?

Physics is concerned with obtaining knowledge about the laws of nature, while psychology is concerned with obtaining the wisdom to maximize success with that knowledge. Knowledge fosters complexity, while wisdom fosters simplicity. Therefore, Lao Zi contracted the complexity of physics towards the simple principle of balance: *"Those who daily pursue the Way (Dao) will contract."* ([9] §48).[35] On the other hand, modern physics expands into an infinite maze of increasingly complexity: *"Those who daily pursue knowledge will expand"* ([9] §48).[36] To understand even a small part of that maze, thinkers are forced to become specialists. That specialization inhibits the communication between the different fields of knowledge, which recreates a linguistic confusion comparable to the biblical Tower of Babel. Ultimately, specialists know everything about nothing. Physicists have to ask: Is the emphasis on empirical research the answer to the manmade extinction problems? Should physicists revaluate holistic thinking? Could humanity solve its extinction problems with the same way of thinking that has created them?

Summary

This paper presents Lao Zi's ancient Theory of Fluctuating Nothingness. It holds that the cosmos consists of the mirror universes of Matter and Antimatter, which fluctuate around Nothingness. The balance between Matter and Antimatter is maintained by the generic force Qi manifesting as specific gravitational, electro-magnetic, and nuclear forces. In Lao Zi's theory: Nothingness is the cosmic constant; Balance is the cosmic principle; and Qi is the cosmic force. Two-and-halve thousand years ago, he was the first to describe gravity, Antimatter, Nothingness, fluctuation, and a zero-sum creation theory

Lao Zi's Theory of Fluctuating Nothingness is supported by modern physics and synthesizes some of its current theories. For example, Einstein's mass-energy equivalent, Schrödinger's quantum-entanglement, and Capra's subatomic-oneness support Lao Zi's idea of cosmic Oneness.

Hawking's zero-energy universe and Tryon's vacuum universe support Lao Zi's concept of cosmic Nothingness.

Lao Zi's theory implies that reactions, entropy, and chaos are restorations of hidden Yin-Yang balances by the force Qi. Hence, his theory synthesizes Newton's Third Law of Motion, Newton's Second Law of Thermodynamics, and Lorenz' Chaos Theory. In Lao Zi's theory, the magnitude of the reaction depends on the characteristics of the particular Yin-Yang balance, rather than on the magnitude of the action. Therefore, his theory synthesizes Newton's idea that Action is Reaction with Lorenz' Butterfly Effect.

Lao Zi's concept of Returning suggests that fluctuation hides the missing universe of Antimatter in time, rather than space. Fluctuation provides also a temporal buffer between the mirror universes of Matter and Antimatter, which prevents their mutual annihilation. Lao Zi's theory changes the history of physics and challenges the purpose, methodology, and philosophical premises of modern physics. It is relative to our era, because it could foster the survival of humanity and research in physics, cosmology; environmental sciences; psychology, psychophysics; and Chinese history and literature.

References

[1] Schade, A., Dictionary Lao Zi's Dao De Jing, 1 ed., vol. 1, Dunedin: nemonik-thinking.org (Printer Amazon), 2018a.

[2] Schade, A., Meta-translation Lao Zi's Dao De Jing (1-37), 3 ed., vol. 2, Dunedin: nemonik-thinking.org (Printer Amazon), 2018b.

[3] Schade, A., Meta-translation Lao Zi's Dao De Jing (38-81), 3 ed., vol. 3, Dunedin: nemonik-thinking.org (Printer Amazon), 2018c.

[4] Schade, A., Lao Zi's Dao De Jing Demystified, 2 ed., vol. 4, Dunedin: nemonik-thinking.org (Printer Amazon), 2017a.

[5] Schade, A., Stunning Revelations about Lao Zi's Dao De Jing, 2 ed., vol. 5, Dunedin: nemonik-thinking.org (Printer Amazon), 2017b.

[6] Zhuang Zi, The Complete Works of Chuang Tzu, New York: Columbia University Press, 1968.

[7] Zhuang Zi, Chuang Tzu, London: Unwin Paperbacks, 1980.

[8] Schade, A., Meta-translation Lao Zi's Dao De Jing (1-37), 3 ed., vol. 2, Dunedin: nemonik-thinking.org (Printer Amazon), 2018.

[9] Schade, A., Meta-translation Lao Zi's Dao De Jing (38-81), 3 ed., vol. 3, Dunedin: nemonik-thinking.org (Printer Amazon), 2018.

[10] Smoot, G. and Davidson, K., Wrinkles in Time. The imprint of Creation., London: Little, Brown and Company, 1993.

[11] Schade, A., Dictionary Lao Zi's Dao De Jing, 1 ed., vol. 1, Dunedin: nemonik-thinking.org (Printer Amazon), 2018.

[12] Newton, Sir Isaac, The Mathematical Principles of Natural Philosophy (Principia), London, 1687.

[13] Hawking, S., A brief History of Time., Great Britain: Bantam Press, 1988.

[14] Einstein, A., "Über die von der molekularkinetischen Theorie der Wärme geforderte Bewegung von in ruhenden Flüssigkeiten suspendierten Teilchen," Annalen der Physik, vol. 322, no. 8, p. 549–560, 1905.

[15] Einstein, A., "Ist die Trägheit eines Körpers von seinem Energieinhalt abhängig?," Annalen der Physik, vol. 18, no. 13, p. 639–641, 1905.

[16] Einstein, A., "Zur Elektrodynamik bewegter Körper," Annalen der Physik, vol. 17, no. 10, p. 891–921., 1905.

[17] Einstein, A., "Über einen die Erzeugung und Verwandlung des Lichtes betreffenden heuristischen Gesichtspunkt," Annalen der Physik, vol. 17, no. 6, p. 132–148, 1905.

[18] Bodanis, D., E=mc2: A Biography of the World's Most Famous Equation (illustrated ed.), Bloomsbury Publishing, 2009.

[19] Dirac, P. A. M., "The Quantum Theory of the Electron," Proceedings of the Royal Society A: Mathematical, Physical and Engineering Sciences, vol. 117, no. 778, p. 610–624, 1928.

[20] Dirac, P. A. M., "A Theory of Electrons and Protons," Proceedings of the Royal Society A: Mathematical, Physical and Engineering Sciences, vol. 126, no. 801, p. 360–365, 1930.

[21] Anderson, C. D., "The Apparent Existence of Easily Deflectable Positives," Science, vol. 1967, no. 76, p. 238–9, 1932.

[22] Anderson, C. D., "The Positive Electron," Physical Review, vol. 43, no. 6, p. 491–494, 1933.

[23] Kumar, M., Quantum, Icon Books, 2009, p. 313.

[24] Born, M., The Born-Einstein Letters 1916-1955, New York: Macmillan Press, 1971.

[25] Einstein A, Podolsky B, and Rosen N., "Can Quantum-Mechanical Description of Physical Reality Be Considered Complete?" Phys. Rev., vol. 47, no. 10, p. 777–780., 1935.

[26] Kocher, C.A. and Commins, E.D., "Polarization Correlation of Photons Emitted in an Atomic Cascade," Physical Review Letters, vol. 18, no. 15, p. 575–577., 1967.

[27] Formaggio, J.A.; Kaiser, D.I.; Murskyj, M.M.; et all., "Violation of the Leggett-Garg inequality in neutrino oscillations," Phys. Rev. Lett., 2016.

[28] Nairz, O.; Arndt, M.; and Anton Zeilinger, A., "Quantum interference experiments with large molecules," American Journal of Physics, vol. 71, p. 319–325, 2003.

[29] Arndt, M.; Nairz, O; Vos-Andreae, J.; et all., "Wave–particle duality of C60 molecules," Nature, vol. 401, no. 6754, p. 680–682, 1999.

[30] Lee, K. C.; Sprague, M. R.; Sussman, B. J.; et all., "Entangling macroscopic diamonds at room temperature," Science, vol. 334, no. 6060, p. 1253–1256., 2011.

[31] Lorenz, E. N., "Deterministic nonperiodic flow," Journal of the Atmospheric Sciences, vol. 20, p. 130–141, 1963.

[32] Capra, F., The Tao of Physics, Shambhala, 1975.

[33] Tryon, E., "Is the Universe a Vacuum Fluctuation?" Nature, vol. 246, p. 396–397, 1973.

[34] Newton, Sir Isaac, Optiks: or, a treatise of the reflextions, refractions, inflextions and colours of light, London: Royal Society, 1704.

[35] Einstein A., Relativity: The Special and General Theory (Translation 1920), New York: H. Holt and Company, 1916.

[36] Schade, A., Stunning Revelations about Lao Zi's Dao De Jing, 2 ed., vol. 5, Dunedin: nemonik-thinking.org (Printer Amazon), 2017.

[37] Schade, A., Lao Zi's Dao De Jing Demystified, 2 ed., vol. 4, Dunedin: nemonik-thinking.org (Printer Amazon), 2017.

Endnotes

[1] 老子 (Lao Zi or Lao Tzu). 道德經 (Dao De Jing or Tao Te Ching). Dao means: Path, Way, Way of Nature, or Physics. De means: Virtue, Way of People, or Psychology. Jing means: Classic. Hence, *Dao De Jing* means: *A Classic about the Physics of Psychology.*

[2] 古 (the Ancients).

[3] 吾言什易知也 什易行也 而人莫之能知也 而莫之能行也 (§70).

[4] Six Chinese versions were analysed and integrated: (Wang Bi, 226-249 AD); (He-Shang Gong, 179-157 BC); (Fu Yi, 555-639 AD); (Mawangdui-A, ~200 BC); (Mawangdui-B, ~200 BC); and (Guodian, ~300 BC).

[5] Nine English translations were analysed and integrated: (Chan Wing-Tsit, 1988); (Cheng Gia-Fu and English, J, 1972); (Henricks, 1993); (Land, 1990); (Lau, 1985); (Lin, J. P., 1977); (Man-ho Kwok; Palmer, M.; & Ramsay, J., 1997); (Waley, 1968); and (Wing, 1986).

[6] For example, Lao Zi refers to *Gravity* as the *Mysterious Female.*

[7] 寻寻 (Infiniteness). 一 (Oneness). 无物 (Nothingness).

[8] 一者其上不谬 其下之不忽 寻寻呵 (§14).

[9] 不可名也 (§14). 道恒无名 (§32 & 37). 道隐无名 (§41).

[10] 复归于无物 是谓无状 之状无物之象 (§14).

[11] 有物昆成先天地生 萧呵寥呵 (§25).

[12] 有物昆成先天地生 萧呵寥呵 独立而不改 可以为(天地)母 吾未知其名字之曰道 (§25).

[13] 吾不知其谁之子也 象帝之先 (§4).

[14] 以知古始 (§14).

[15] 一生二 (§42).

[16] 天下之物生于有 有生于无 (§40). The Mawangdui texts (A & B) were buried from 168 BCE to 1973 CE. In both versions, the part traditionally known as chapter 40 precedes immediately chapter 42 [11]. As a result; two sentences about Matter and Antimatter precede

immediately Lao Zi's statements about the formation of the universe. This supports the notion that the Two mean: Matter and Antimatter, rather than Yang and Yin.

[17] 归, 复, 反, or 返 (return).

[18] 复归于无物 (§14).

[19] 万物傍作 吾以观其复也 夫物秐秐 各复归于其根 归根曰静 静是谓复命 复命常也 (§ 16).

[20] 逝曰远 远曰反 (§25).

[21] 反也者道之动也 (§40).

[22] Lao Zi did not name his theory.

[23] 阴 (Yin); 阳 (Yang).

[24] 万物负阴 而抱阳 (§42).

[25] 冲气以为和 (§42).

[26] 谷神不死是谓玄牝 (§6).

[27] 不死 (immortal or eternal). 谷 (valley or downwards). 神 (spirit or force).

[28] 中有物呵 幽呵冥呵 其中有精呵 其精什真 (§21).

[29] 致虚极也 守静笃也 (§16).

[30] 求以得有罪以免与 (§62)

[31] 焉人法地 地法天 天法道 道法自然 (§25).

[32] 不道早已 (§30 & 55).

[33] 夫天道无亲 恒与善人 (§79).

[34] 善人之保也 (§62).

[35] 为学者日益 (§48).

[36] 为道者日损 (§48).

6.3 Ancient Theory Challenges Modern Psychology

Dr Auke J. Schade

Unpublished manuscript 3 June 2021

Abstract

About two-and-halve thousand years ago, Lao Zi wrote Dao De Jing, which means: A Classic about the Physics of Psychology. His aim is to maximize success, which he defined as to obtain what you seek and escape what you suffer. Unfortunately, scholars have mystified his book by explaining his scientific theories in supernatural terminology. Manmade extinction problems, such as nuclear war and climate change, show that humanity suffers from systemic cognitive failure. Those problems cannot be solved with the same way of thinking that has created them. Humanity needs a valid and reliable theory for problem solving. Lao Zi's Theory of Fluctuating Nothingness provides a rationale for modern physics. His theory holds that balance is the cosmic principle, while the force Qi maintains and restores that balance. Qi comprises the gravitational, electro-magnetic, and nuclear forces. People maximize their success by aligning with the unstoppable Qi. Therefore, physics is the only valid and reliable basis for psychology. The aim of this study is to evaluate Lao Zi's Cognitive Criteria for Problem-solving. Those criteria include: focus, balance, efficiency, and proactivity.

Dao De Jing

The Chinese philosopher Lao Zi (~570-490 BCE) wrote Dao De Jing.[1] Dao means: Path, Way, Way of Nature, or Physics. De means: Virtue, Way of People, or Psychology. Jing means: Classic. Hence, Dao De Jing means: A Classic about the Physics of Psychology [1] [2] [3] [4] [5]. The theories recorded in Dao De Jing were developed by a civilization that Lao Zi called: The Ancients ([2] §15 & 22); ([3] §62, 65 & 68).[2] Zhuang Zi (~369-286 BCE) and subsequent scholars have mystified Dao De Jing with their attempts to explain Lao Zi's scientific theories in supernatural terminology [6] [7] [8] [9]. It is the aim of this paper to demystify Lao Zi's Dao De Jing and show the importance of that ancient manuscript for cognitive psychology.

A seven-stage translation-interpretation process was applied to demystify Lao Zi's Dao De Jing. (I) A special dictionary was compiled containing the exhaustive English meanings for each pictograph mentioned in each of the six significant Chinese versions of Dao De Jing [1].[3] (II) The translations from Lao Zi's archaic Chinese pictographs to modern ones were validated by integrating those six Chinese versions [2] [3]. (III) The translation from modern Chinese to English was supported by comparing the integrated Chinese text with nine significant English translations. To determine the final meaning of each pictograph, the consistency of that meaning was maximized across the entire Dao De Jing [2] [3]. (IV) To clarify Lao Zi's theory, the holistic format of Dao De Jing was reorganized into a rational format organized by topic [4]. (V) Lao Zi's ancient jargon was updated to modern physics [4].[5] (VI) Lao Zi's theory of physics was reconstructed and named: Theory of Fluctuating Nothingness [5]. (VII) Lao Zi's theory of psychology was reconstructed and named: Cognitive Criteria for Problem Solving [5].

Dao or Physics

Lao Zi's Theory of Fluctuating Nothingness holds that Nothingness is the eternal constant that was divided into mirror universes of Matter and Antimatter. The constancy of Nothingness forces Matter and Antimatter to maintain a balance of equal magnitudes. Therefore, balance is the principle of physics. Lao Zi's theory of balance is supported by the modern balance or equal sign (=) that is fundamental for all sciences, arithmetic, mathematics, accountancy, and logic.

The cosmos contains all Matter and Antimatter, while the universe contains only Matter. The cosmic balance between Matter and Antimatter underlies the multitude of Yin-Yang balances constituting our universe. Yin means literally: *shady side of the hill.*[6] However, it represents a cluster of negative concepts [1]. On the other hand, Yang means literally: *sunny side of the hill.*[7] However, it represents a cluster of positive concepts [1]. Hence, Yin-Yang balances comprise polarities such as: black-white; weak-strong; back-front; bottom-top; female-male; passive-active; etc. The Yin-Yang balances are maintained by the generic force Qi in its manifestations as specific gravitational, electro-magnetic, and nuclear forces. Yin-Yang balances interact as domino chains.

Lao Zi's ancient theory provides the rationale for several concepts of modern physics such as: gravity, entropy, motion, mass-energy equivalence, antimatter, quantum-entanglement, chaos, subatomic-oneness, zero-energy universe, zero-sum creation, and vacuum fluctuation [5] [10].

De or Psychology

Cognition includes thinking and memory. Thinking is an associative cognitive process that reorganizes available information in order to generate new information. On the other hand, memory is an associative cognitive process that stores, maintains, and recalls information in order to preserve that information across space and time. The literature contains several discussions concerning the information flow through the cognitive architecture [11] [12] [13] [14]. In addition, the present study presents Lao Zi's cognitive criteria, which aim to manage the gates of that information flow in order to maximize success. Lao Zi defined success as: *"To obtain what you seek and to escape what you suffer"* ([3] §62).[8]

Unfortunately, reality shows that humanity creates a growing host of manmade extinction problems such as: climate change, dwindling resources, overpopulation, pollution, and technological warfare. These problems suggest that our species suffers from systemic cognitive failure. Each solution creates a multitude of new problems. That is no surprise, because our way of Rational Thinking was mainly developed by the Greek philosopher Aristotle (384–322 BCE) and was never properly updated. In addition, technological progress has changed the static reality of ancient Greece into the modern hyper-dynamic one. Therefore, static thinking has turned from an asset into a liability.

The Greek way of thinking that is propagated by the educational system is predominantly based on inductive and deductive thinking. Both methods require complete, valid, and reliable facts. The search for such facts drives thinkers into a complex and infinite maze of details. To understand even a small part of that maze, thinkers are forced to become specialists. That specialization breaks down the communication between the different fields of knowledge, which recreates a linguistic confusion comparable to the biblical Tower of Babel. Ultimately, specialists know everything about nothing.

Our perception of reality is biased towards movement. Lao Zi's principle of balance and the rebalancing force Qi are ignored when they are at rest. Instead, we clearly perceive the roaring avalanche restoring the gravitational balance; the blinding lightning restoring the electro-magnetic balance; and the devastating explosion restoring the nuclear balance. Hidden gravitational, electro-magnetic, and nuclear forces move matter around. Therefore, we perceive an external reality of chaos and entropy. Nevertheless, that apparent turmoil is the result of the underlying balance. Hence, Lao Zi's principle of balance is the only constant in the external reality: *"Knowing this constant (balance) is brilliant. Not acknowledging this constant is arrogant. Arrogance causes misfortune* ([2] §16). *Knowing harmony (balance) is called the constant. Knowing the constant is called brilliance"* ([3] §55).[9] Lao Zi's principle of balance and the rebalancing force Qi simplifies the complex reality. Therefore, those ancient concepts should be taken into account for modern problem-solving.

Unfortunately, cognitive improvement is inhibited by educational conditioning, institutionalization, groupthink, and cognitive dissonance. Nevertheless, we cannot solve problems with the same way of thinking that has created them. We have to improve or face the dire consequences: *"What is not the Way (Physics) will perish soon"* ([2] §30).[10] Static thinking and opposing improvement is dangerous: *"The hard and strong are called companions of death. The soft and weak are the companions of life... A strong tree will be broken"* ([3] §76).[11] Instead, perfection is the ability to adjust to the dynamic reality: *"Bend then be preserved"* ([2] §22).[12]

Decision-making Criteria

Lao Zi's Theory of Fluctuating Nothingness holds that the division of the constant cosmic Nothingness in equal opposites fosters the principle of cosmic balance [5] [10]. That idea is supported by the equal sign (=), which is the basis of all modern sciences. Hence, if Lao Zi's theory would be rejected, then his principle of cosmic balance would be still valid. Figure 1 shows that that cosmic balance is maintained by the force Qi. Within that dynamic reality, there are four exhaustive cognitive criteria for maximizing success:

1. Focus.

2. Balance.

3. Efficient.

4. Proactive.

Together, those four cognitive criteria provide the acid-test for problem-solving. They foster holistic thinking that evaluates the dynamic relationships between reality, goal, problem, solution, decision, and reaction. After sufficient practice; consciously contemplating any of those criteria will prompt the subconscious to maximize success by recalling and processing information associated with that criterion.

Figure 1: Lao Zi's De — *Cognitive Criteria Theory of Problem-solving*

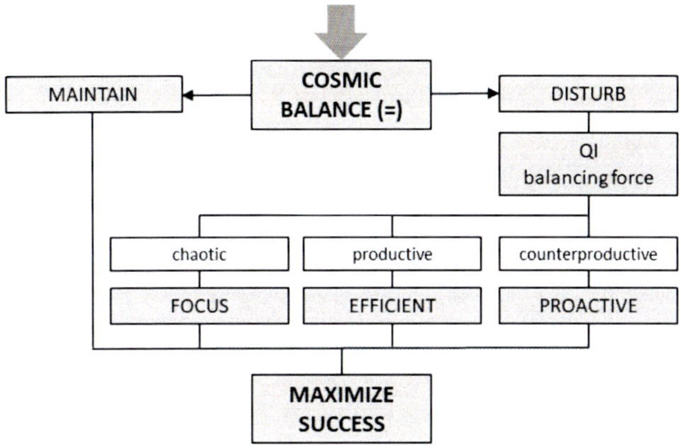

Focus

Contemplating consciously the word *'Focus'* prompts the subconscious to concentrate on maximizing your success. You can only maximize your success if you focus on the Way, Way of Nature, or Physics: *"The greatest virtue is following only the Way"* ([2] §21). *"Have a long life through a lasting regard for the Way"* ([3] §59).[13] However, Lao Zi's Theory of Fluctuating Nothingness holds that movement, chaos, and entropy are detectable restorations of the Yin-Yang balances by Qi [5]

[10]. That turmoil could distract you from your goals: *"The bright Way seems to be obscure; the Way forward seems to be backwards; and the smooth Way seems to be rough"* ([3] §41).[14] Incompetent people fail, because they focus on obtaining what they should escape, while escaping what they should obtain: *"What is not the Way (Physics) will perish soon"* ([3] §55).[15] Therefore, make sure that you know what to obtain and what to escape.

The universe is fair—if you ask for nothing, then you will get nothing. Before you can hit the target, you must select one. Do not let other people replace your goals with theirs. If a problem does not affect your success, then it is not your problem. Balance is the cosmic principle and, therefore, sages balance also their focus. If you focus exclusively on obtaining what you seek, then you might not escape what you suffer. A mouse obtains the cheese, but does not escape the deadly trap. In contrast, if you focus exclusively on escaping what you suffer, then you might not obtain what you seek. A deer escapes the noise of the hunting dogs without obtaining safety. Instead, it will be killed by the waiting hunters.

A strong focus can be as counterproductive as a weak one. It could be so strong that it inhibits the consideration of opportunities that are more productive. Focus fosters static thinking that will fail sooner or later in our hyperdynamic reality. Therefore, your focus has to be updated to the continuously changing situation. Freeze and unfreeze your focus at the right time: *"In action, the goodness is timing"* ([2] §8).16 Hang on tightly, but let go lightly.

Lao Zi explained that the immutable laws of nature force people to focus on their success if they want to maximize their success. The concept of focus or attention was rediscovered during the ninetieth century [15] [16].

Balance

Contemplating consciously the word *'Balance'* prompts the subconscious to maximize success by maintaining the balance in order to prevent a counterproductive rebalancing by Qi. Lao Zi's Theory of Fluctuating Nothingness holds that everything consists of interacting Yin-Yang balances that are maintained by Qi: *"All-things carry Yin on their back and carry Yang in their arms. Any difference creates vital energy (Qi) that restores their harmony"* ([3] §42).[17] Hence, every action evokes an opposing reaction.[18] Therefore, Lao Zi cautioned: *"Sages manage their affairs without action* ([2] §2). *When walking on the Great Road (Applying Physics), the only thing scary is action* (§53). *All sages will avoid great actions* (§63). *Those who act will fail"* ([3] §64).[19]

Extremes disturb the Yin-Yang balances. Therefore, Lao Zi advised: *"Concentrate on removing extremes. Nurture tranquillity (balance) diligently* (§16). *Sages reject extremes, reject grandeur, and reject extravagance* ([2] §29). *No greater suffering than having extreme desires. No greater misfortune than not knowing what is enough. No misfortune is more disastrous than the desire to accumulate. Therefore, know that enough is enough and there will be always enough* (§46). *I have always three treasures that I keep and protect. The first one is called compassion. The second one is called frugality. The third one is called humbleness"* ([3] §67).[20]

Lao Zi warned: *"The more people know, the stranger the things they begin to develop"* ([3] §57).[21] In accord, technocrats intentionally disturb the Yin-Yang balances in order to evoke Qi and use it for heating, transport, communication, manufacturing, warfare, etc. Nevertheless, disturbing a balance will always initiate a chain reaction of rebalancing forces. Lao Zi warned us for the complexity and unpredictability of such restorations: *"Those who pretend to know the future are the fruitless flowers of the Way (Physics) and the chiefs of fools"* ([3] §38).[22] In accord, reality shows that even small actions had huge and unpredictable consequences. For example, releasing relatively small amounts of allegedly safe CFC, CO2, and DDT have created skin cancer, climate change, and annihilated entire food chains. We create such manmade extinction problems, because we apply knowledge without wisdom.

Many scholars have emphasized the importance of maintaining the balance and alternatively named it: Moderation, Temperance, Doctrine of the Mean, Golden Path, etc. [17] [18] [19] [20] [21]. However, only Lao Zi's Theory of Fluctuating Nothingness explains why and how the immutable laws of nature force people to maintain the balance if they want to maximize their success.

Efficient

Contemplating consciously the word *'Efficient'* prompts the subconscious to maximize success with a minimum of effort and resources by aligning with the rebalancing Qi. Lao Zi referred to efficiency as: Action-without-Action, Action-without-Effort, or Non-action: *"Sages manage their affairs with Non-action* (§2). *"Act with Non-action, then there will be no anarchy* ([2] §3). *Superior virtue uses Non-action and there is no action used* (§38). *Act with Non-action* (§63). *Sages will use Non-action. Therefore, they will not fail"* ([3] §64).[23] Resources are always limited, independent of their initial number or amount. If used inefficiently, they will be depleted sooner or later. Even clean air and fresh water have become scarce. Therefore, Lao Zi wrote: *"Those who are frugal can be generous"* ([3] §67).[24]

Rowing a boat requires effort and is exhausting. Therefore, sages act like sailors who set their sails to the wind, which will bring them to their destiny without much effort: *"Sages can achieve greatness, because they do not act great"* ([2] §34).[25] Any effort indicates inefficiency that is caused by opposing Qi. Lao Zi wrote: *"If one wants to take the world, then one should always use no effort. When effort is needed, then there is never enough to take the world"* (§48). *"Use no effort when taking the world" (§57). "Work without effort"* ([3] §63).[26] That does not mean that sages never act. They continuously position themselves and follow the way of least resistance: *"Act, but do not rely on it* (§2 & 51). *Sages act, but do not rely on it"* ([3] §77).[27]

Efficiency could inhibit your focus on success, because going with the flow of the river will never bring you to the top of the mountain. Furthermore, efficiency means doing things right, but that does not necessarily means doing the right things. Therefore, the results of efficiency can be inhumane, unkind, controlling, and bureaucratic. Such uncaring efficiency could alienate you from other people, which weakens your psychological and physical support network. Hence, sages do not cut the branch they are sitting on: *"Those who use compassion to attack will triumph. Those who use it to defend will stand firm"* ([3] §67).[28]

Lao Zi explained that the immutable laws of nature force people to be efficient if they want to maximize their success. The concept of efficiency was rediscovered by the Efficiency Movement during the early twentieth century [22] [23] [24].

Proactive

Contemplating consciously the word *'Proactive'* prompts the subconscious to maximize success by restoring the balance before the rebalancing Qi gains a counterproductive momentum. Proactivity is efficient, because it preserves resources for future problems. Therefore, Lao Zi suggested: *"Make the large small and the many few... Pursue the difficult, while it is easy. Act large, while it is still small. The world's most difficult things arise from the easiest. The world's largest things arise from the smallest* (§63). *That what is at rest is easy to hold. That what is not manifest is easy to plan. That what is fragile is easy to break. That what is small is easy to scatter. Act when it has not happened yet. Control it when it is not chaotic yet. A tree that takes both arms to embrace grows from a little cutting"* ([3] §64).[29]

Prevention is the best cure, but correct timing is crucial for any action: *"In action, the goodness is timing"* ([2] §8).[30] For example, we might have to inhibit global warming. If we wait too long, then it might destroy the resources required for climate actions. However, proactivity requires a correct prediction of the complex reality. Lao Zi warned: *"Those who pretend to know the future are the fruitless flowers of the Way (Physics) and the chiefs of fools"* ([3] §38).[31] Hence, proactive climate action could be in reality a counterproductive rash action. We might lower the atmospheric temperature, while nature moves us at the same time into the next ice-age. Such actions could destroy opportunities and create threats: *"Those who make rash promises are certainly difficult to trust. Those who regard everything as easy will have certainly many difficulties"* ([3] §63).[32]

Lao Zi explained that the immutable laws of nature force people to be proactive if they want to maximize their success. The concept of proactivity was rediscovered in the twentieth century [25] [26].

Implications

Lao Zi's *'Cognitive Criteria Theory'* holds that the cosmic principle of balance provides four criteria for problem-solving. His theory could foster empirical research in several fields including: education, business management, and cognitive- and neuro-psychology. For example, the following hypotheses could be tested concerning subjects trained in problem-solving with Lao Zi's four criteria:

1. Trained subjects solve practical problems significantly faster and better than untrained subjects.

2. Trained subjects will show a significant improvement in creativity and self-confidence.

3. Brain scans of trained subjects are significantly different from scans of untrained subjects.

Recently, Lao Zi's Cognitive Criteria Theory was extended to Nemonik Thinking [27] [28] [29]. Nemonik Thinking is the operating system for the brain that maximizes success. It incorporates Lao Zi's four cognitive criteria and adds seventeen new ones. Those twenty-one cognitive criteria are called nemoniks, which provide an exhaustive checklist for problem-solving. Consciously contemplating a nemonik activates subconscious guided-association, which recalls and processes information associated with that nemonik within the context of reality. Nemonik Thinking aims to make people the best thinkers they can be. Such thinkers might be able to prevent and solve manmade extinction problems.

Summary

Lao Zi proposed that physics is the only valid and reliable basis for psychology. His Theory of Fluctuating Nothingness holds that balance is the cosmic principle and Qi is the balancing force. In accord with Behaviourism, Lao Zi's Cognitive Criteria Theory holds that Qi conditions people to align with the cosmic balance. Qi rewards them by presenting what they seek and withholding what they suffer. On the other hand, Qi punishes them by withholding what they seek and presenting what they suffer. That conditioning fosters four natural principles for problem-solving including: Focus, Maintaining balance, Efficient, and Proactive.

By continuously removing imbalances, the force Qi creates apparent chaos and entropy in the external reality. That turmoil is likely distracting people from their goals. Therefore, sages focus on maximizing their success. Qi will always maintain and restore the balance. Therefore, sages will maintain the balance by avoiding actions that evoke counterproductive reactions. If the balance is disturbed, then the rebalancing force Qi is unstoppable. In that case, sages align with the force Qi. Sages are efficient and take advantage of the free energy provided by Qi. If an imbalance could evoke counterproductive reactions from Qi, then sages will be proactive. They will restore that balance before Qi gains momentum.

The growing host of manmade extinction problems shows that humanity is not focusing on maximizing success. Instead of maintaining balances, technocrats purposely disturb them evoking unpredictable and counterproductive reactions. Furthermore, humanity is inefficient, because it wastes time, effort, and resources in opposing the counterproductive rebalances by Qi. Last but not least, humanity's global response is retroactive, rather than proactive. Unfortunately, cognitive improvement is inhibited by the institutionalized way of thinking.

Lao Zi's Cognitive Criteria Theory could foster research in all fields of knowledge including: education, business management, cognitive- and neuro-psychology, etc. Recently, his theory was extended to Nemonik Thinking, which comprises an exhaustive checklist of twenty-one cognitive criteria for problem-solving. Therefore, incorporating Nemonik Thinking in the educational curriculum might prevent and solve manmade extinction problems.

References

[1] Schade, A., Dictionary Lao Zi's Dao De Jing, 1 ed., vol. 1, Dunedin: nemonik-thinking.org (Printer Amazon), 2018.

[2] Schade, A., Meta-translation Lao Zi's Dao De Jing (1-37), 3 ed., vol. 2, Dunedin: nemonik-thinking.org (Printer Amazon), 2018.

[3] Schade, A., Meta-translation Lao Zi's Dao De Jing (38-81), 3 ed., vol. 3, Dunedin: nemonik-thinking.org (Printer Amazon), 2018.

[4] Schade, A., Lao Zi's Dao De Jing Demystified, 2 ed., vol. 4, Dunedin: nemonik-thinking.org (Printer Amazon), 2017.

[5] Schade, A., Stunning Revelations about Lao Zi's Dao De Jing, 2 ed., vol. 5, Dunedin: nemonik-thinking.org (Printer Amazon), 2017.

[6] Zhuang Zi, The Complete Works of Chuang Tzu, New York: Columbia University Press, 1968.

[7] Zhuang Zi, Chuang Tzu, London: Unwin Paperbacks, 1980.

[8] Palmer, M.; Kwok, M.; and O'Brien, J., The Eight Immortals of Taoism, Sydney: Random Century Group Australia (Pty) Ltd., 1990.

[9] Yu-lan, F., A History of Chinese Philosophy, USA: Princeton University Press, 1983.

[10] Schade, A., "Ancient Theory Challenges Modern Physics," Journal ### (submission in progress), vol. ###, no. ###, p. ###, 2021.

[11] Anderson, J. R., The architecture of cognition, Cambridge: Harvard University Press, 1983.

[12] Fodor, J. A., The modularity of the mind, New York: MIT Press, 1975.

[13] McCloskey, M., "Networks and theories. The place of connectionism in cognitive science," Psychological Science, vol. 2, pp. 387-395, 1991.

[14] Stillings, N. A.; Weisler, S. E.; Chase, C. H.; et al., Cognitive Science. An introduction., A Bradford Book, 1995.

[15] James, W., The Principles of Psychology, New York: Henry Holt, 1890.

[16] Anderson, J. R., Cognitive Psychology and Its Implications, Worth Publishers, 2004.

[17] Dhand, A., "The dharma of ethics, the ethics of dharma: Quizzing the ideals of Hinduism," Journal of Religious Ethics, vol. 30, no. 3, pp. 347-372, 2002.

[18] Hashim, K. M., The Middle Path of Moderation in Islam: The Qur'ānic Principle of Wasaṭiyyah, Oxford: Oxford University Press, 2015.

[19] Omar, H. A., Moderation in Islam, United Printing Publishing and Distributing, 1999.

[20] Peterson, C. and Seligman, M. E. P., Character strengths and virtues: A handbook and classification, New York: Oxford University Press, 2004.

[21] Nāgārjuna, The philosophy of the middle way = Mūlamadhyamakakārikā, D. Kalupahana, Ed., Albany, N.Y.: State University of New York Press, 1986.

[22] White, A., Efficiency and empire, London: Methuen & Co, 1901.

[23] Spillman, W. J., "The Efficiency Movement in Its Relation to Agriculture," Annals of the American Academy of Political and Social Science, vol. 59, no. May, p. 65–76, 1915.

[24] Heath, J., The Efficient Society: Why Canada Is as Close to Utopia as It Gets., Toronto: Penguin Canada, 2002.

[25] Whiteley, P. L. and Blankfort, G., "The Influence of Certain Prior Conditions Upon Learning," Journal of Experimental Psychology, vol. 16, p. 843–851, 1933.

[26] Crant, J. M., "Proactive Behavior in Organizations," Journal of Management, vol. 26, no. 3, p. 435, 2000.

[27] Schade, A., Think Smarter with Nemonik Thinking, Dunedin: nemonik-thinking.org (Printer Amazon), 2016.

[28] Schade, A., Dictionary Nemonik Thinking, Dunedin: nemonik-thinking.org (Printer Amazon), 2016.

[29] Schade, A., Glossary nemonik Thinking, Dunedin: nemonik-thinking.org (Printer Amazon), 2016.

Endnotes

1 老子 (Lao Zi or Lao Tzu). 道德經 (Dao De Jing or Tao Te Ching).

2 古 (the Ancients).

3 Six Chinese versions were analysed and integrated: (Wang Bi, 226-249 AD); (He-Shang Gong, 179-157 BC); (Fu Yi, 555-639 AD); (Mawangdui-A, ~200 BC); (Mawangdui-B, ~200 BC); and (Guodian, ~300 BC).

4 Nine English translations were analysed and integrated: (Chan Wing-Tsit, 1988); (Cheng Gia-Fu and English, J, 1972); (Henricks, 1993); (Land, 1990); (Lau, 1985); (Lin, J. P., 1977); (Man-ho Kwok; Palmer, M.; & Ramsay, J., 1997); (Waley, 1968); and (Wing, 1986).

5 For example, Lao Zi refers to Gravity as the Mysterious Female.

6 阴 (Yin).

7 阳 (Yang).

8 求以得有罪以免与 (§62).

9 知常明也 不知常妄 妄作凶 (§16). 知和曰常 知常曰明 (§55).

10 不道早已 (§30).

11 木强则折 (§76).

12 曲则全 (§22).

13 孔德之容唯道是从 (§21). 长生久视之道也 (§59).

14 明道如昧 进道若退 夷道如类 (§41).

15 不道早已 (§55).

16 动善时 (§8).

17 万物负阴而抱阳 冲气以为和 (§42).

18 In Newton's sterile reality, each action evokes a reaction of equal magnitude and opposite direction. However, Lao Zi's practical reality consist of a multitude of Yin-Yang

balances interacting as domino-chains. Therefore, small actions could evoke gigantic reactions.

19 圣人处无为之事 (§2). 行于大道唯施是畏 (§53). 圣人终不为大 (§63). 为之者败之 (§64).

20 致虚极也 守静笃也 (§16). 圣人去什 去大去诸 (§29). 罪莫大于什欲 祸莫大于不知足 咎莫憯于欲得 故知足之为足恒足矣 (§46). 我恒有三宝持而保之 一曰慈 二曰俭 三曰不敢为天下先 (§67).

21 人多知而奇物滋起 (§57).

22 前识者道之华也而愚之首也 (§38).

23 圣人处无为之事 (§2). 为无为则无不治矣 (§3). 上德无为而无以为也 (§38). 为无为 (§63). 圣人无为也 故无败也 (§64).

24 俭故能广 (§67).

25 圣人能成大也以其不为大也 (§34).

26 取天下也恒以无事 及其有事也不足以取天下矣 (§48). 以无事取天下 (§57). 事无事 (§63).

27 为而弗恃也 (§51). 圣人为而弗恃 (§77).

28 夫慈以战则胜 以守则固 (§67).

29 大小多少 图难乎于其易 为大乎于其细也 天下之难作于易 天下之大作于细 (§63). 其安也易持也 其未兆也易谋也 其脆也易破也 其微也易散也 为之于其未有也治之于其未乱也 合抱之木 (§64).

30 动善时 (§8).

31 前识者道之华也而愚之首也 (§38).

32 夫轻诺必寡信 多易必多难 (§63).

6.4 The Cognitive Revolution

Dr Auke J. Schade

Unpublished manuscript 15 August 2021

Abstract

For many millennia, humanity has lived in a relatively static reality. However, during the last few centuries, education has fostered the science and technology that has turned that static reality into the modern hyperdynamic one. Unfortunately, the cognitive adjustment to that environmental dynamism lags behind. As a result, people have become static thinkers in a hyperdynamic reality. That contradiction has created manmade extinction problems such as nuclear war and climate change. Such threats of self-destruction are scientific evidence that humanity suffers from systemic cognitive failure. Despite the advantages of the educational system, it is complicit in those threats. Science and technology might not be the answer. Although computers are fast, they are unlikely to solve those complex extinction problems, because they are simplified replicas of the human mind. Instead, the educational system should adopt a way of thinking that fits the hyperdynamic reality. Based on ideas of Lao Zi, Aristotle, and Hegel; the Cognitive Accelerator is introduced in order to foster the cognitive revolution from Static to Dynamic Thinking. Humanity needs a revolution, because it is rapidly running out of time.

Systemic Cognitive Failure

Cognition includes memory and thinking. Memory is a self-organising associative cognitive system that stores, maintains, and recalls information in order to preserve that information across space and time. On the other hand, thinking is a self-organizing cognitive system that processes available information in order to generate new information.

The Greek philosopher Aristotle (384-322 BC) introduced validity rules for logic. Those rules were compiled by the Peripatetics in the Organon, which means the Instrument [1]. Logic directs our thinking from known facts to an irrefutable conclusion. A fact is a valid and reliable description of reality that is adequately supported by observation and/or logic. The conclusion becomes a new fact and so on.[1]

As shown in Figure I: Rational Thinking is a conscious way of thinking that submits facts to logic in order to find the truth. It includes OBJECTIVE and COLLECTIVE mindmodes. A mindmode is a way of thinking that deals with a specific aspect of reality. OBJECTIVE refers to the unchangeable laws of nature, while COLLECTIVE refers to the changeable rules made by groups of people. Logical-, analytical-, and critical-thinking are components of Rational Thinking [2].

In contrast to Rational Thinking, Affectorial Thinking is a subconscious way of thinking that includes the CREATIVE and REACTIVE mindmodes. Creative is associated with the future, while reactive is associated with the past. Affectorial Thinking produces affecters, which are signals from the subconscious aiming to inform and influence the conscious. Affecters include ideas, intuitions, emotions, and skills [2].

Figure I: Diagram of the Mind.

MIND			
CONSCIOUS		SUBCONSCIOUS	
Rational Thinking (facts)		Affectorial Thinking (<< affecters)	
Nature	People	Future	Past
Unchangeable laws	Changeable rules	Ideas	Memory
Objective mindmode	**Collective mindmode**	**Creative mindmode**	**Reactive mindmode**

The Chinese philosopher Lao Zi (~570-490 BCE) wrote the book Dao De Jing, which means: A Classic about the Physics of Psychology [3] [4] [5] [6] [7].[2] However, his theories about physics and psychology were developed by a civilization he called: The Ancients ([4] §15 & 22); ([5] §62, 65 & 68).[3] Lao Zi's warning supports the sophistication of those theories: *"My words are very easy to understand and very easy to apply. Yet, people cannot understand them and they cannot apply them"* ([5] §70).[4]

Zhuang Zi (~369-286 BCE) and subsequent scholars mystified unintentionally Dao De Jing with their attempts to explain Lao Zi's theory of physics in super-natural terminology [8] [9]. Physics needed to recover, before the knowledge of the Ancients could be recognized. In accord, modern physics is the scientific basis for the presented demystification of Lao Zi's Dao De Jing [6] [7].

A translation-interpretation process of several distinct stages was applied to demystify Lao Zi's Dao De Jing. (I) A special dictionary was compiled containing the exhaustive English meanings for each pictograph mentioned in six reputable Chinese versions of Dao De Jing [3].[5] (II) The translations from Lao Zi's archaic Chinese pictographs to modern ones were validated by integrating those six Chinese versions [4] [5]. (III) The translation from modern Chinese to English was supported by comparing the integrated Chinese text with nine reputable English translations.[6] To determine the final meaning of each pictograph, the consistency of that meaning was maximized across the entire Dao De Jing [4] [5]. (IV) To clarify Lao Zi's theories, the holistic format of Dao De Jing was reorganized into a rational format organized by topic [6].[7] (V) Lao Zi's ancient jargon was updated to the jargon of modern physics [6]. (VI) Lao Zi's theory of physics was reconstructed and named: Theory of Fluctuating Nothingness [7]. (VII) Lao Zi's theory of psychology was reconstructed and named: Cognitive Criteria Theory [7].

Lao Zi's Cognitive Criteria Theory holds that physics is the only valid and reliable basis for psychology. People are conditioned by the immutable laws of physics as rats in a Skinner box. Consequently, reality conditions humans to adjust to reality. In accord, Lao Zi recommended: *"The greatest virtue is following only physics* [4] (§21).[8] *Not losing life to disaster is called following physics* [5] (§52).[9] *What is not physics will perish soon"* [5] (§55).[10] Lao Zi's ideas about conditioning are supported by the school of Behavioural Psychology [10].

Compared to other species; humans have little physical strength, fangs, or claws. Their mobility and sensory perception are also severely limited. To succeed in the struggle for survival, the environmental pressure conditioned them to rely on their intelligence. Until the beginning of the technological revolutions, people have lived in a relatively static reality. That reality was predictable, because events were repeated over and over again. As a result, that static reality conditioned humans to become static thinkers. Changing a successful solution was counterproductive. Therefore, reality conditioned them to foster memory, reactive thinking, and cognitive inflexibility. That way of thinking was so successful that humans became the apex-predators driving other species towards extinction.

Since the Renaissance, Aristotelian logic has fostered science, while science has fostered the technology that drove the advantageous industrial, biotechnical, and informational revolutions.[11] Those technological revolutions turned the archaic static reality into the modern hyperdynamic one. That new dynamism required intuition, creative thinking, and cognitive flexibility.

The Aristotelian cognitive transformation from dogmatic to Rational Thinking, was initially inhibited by widespread dogmatism, groupthink, and cognitive dissonance. For example, the Catholic Church placed the polymath Galileo Galilei (1564-1642) under house arrest for applying Rational Thinking to astronomy [11]. Today, dogmatism, groupthink, and cognitive dissonance inhibit the transformation from Static to Dynamic Thinking. As a result, humanity's cognitive evolution lags behind their technological revolutions.

Humans became static thinkers in a hyperdynamic reality, which caused humanity's systemic cognitive failure. That failure creates and maintains manmade extinction problems such as nuclear war and climate change.

Lao Zi's ancient theories are crucial for modern empirical research, because empirical research without a theory is like a ship without a rudder. His theory of physics holds that balance is the principle of physics. That idea is supported by Sir Isaac Newton's Third Law of Motion, predicting that each action evokes a reaction of equal magnitude and opposite direction [12]. Newton's reaction is a manifestation of the force that restores Lao Zi's principle of balance. In addition, Lao Zi's cosmic principle of balance is supported by the equal sign (=), which is fundamental to all modern sciences, arithmetic, mathematics, accountancy, and logic.

Lao Zi describes reality as a balanced system in which entropy and chaos are manifestations of forces restoring that balance [7]. Therefore, Lao Zi wrote: *"Knowing balance is called the constant. Knowing the constant is called brilliance"* [5] (§55). On the other hand, modern physics describes reality in terms of entropy and chaos [12] [13]. In accord, modern thinkers focus on the natural effects, rather than the cause. Modern thinkers perceive the punishments and rewards that they receive from the hyperdynamic reality as random events. There is no observable cause and effect. Therefore, reality cannot condition the modern thinkers to adjust to reality. That ignorance causes the systemic cognitive failure of humanity.

Static thinkers are cognitively inflexible, which forces them to adjust the reality to their way of thinking. Therefore, they seek control and use win-lose strategies, which result in counterproductive reactions, conflicts, and aggression. Modern static thinkers are described in Goethe's poem: *The Sorcerer's Apprentice* [14]. They apply knowledge without wisdom. Wisdom is the long-term productive application of knowledge. Each solution evokes a multitude of increasingly larger problems. Static thinkers cannot solve those problems with the same way of thinking that creates them. They have to adjust to the hyperdynamic reality, because Lao Zi's theory of physics predicts that the natural balance will be restored with or without them.

Educational Complicity

The triangle of knowledge comprises education, science, and technology. Education fosters science, while science fosters technology. Unfortunately, technology has created the manmade extinction problems that humanity is facing today. Those threats would not exist without education. Therefore, education is complicit in humanity's self-destruction. Education has to accept that responsibility. It is the only institution that could initiate large scale cognitive improvements. Such improvements are crucial, because humanity is rapidly running out of time [15].

Lao Zi defined success as: *"To obtain what you seek and escape what you suffer"* [5] (§62).[12] In accord, the Utopian aim of education is the development of individual skills. However, the educational system is controlled by the state. Therefore, the mundane aim of the educational system is to provide a curriculum that fits the requirements of the state. Education evaluates each student's understanding of that curriculum in order to assign them to their most productive positions in the socio-economic hierarchy. That shows that educational success is measured by obtaining socio-economic positions, rather than achieving individual goals.

Conditioning is a psychological tool that reinforces a target behaviour with rewards and punishments [10]. In accord, the educational system conditions students with praise, grading, exam results, diplomas, and degrees in order to internalize its goals and way of thinking. That conditioning is highly effective—because it starts at a young age, it is long-term, compulsory, controls the study conditions, and inhibits outside influence. Above all, the educational system determines the future socio-economic status of the student. Hence, students are conditioned like Pavlov's dogs. They either comply or perish. Dogmatism, groupthink, and cognitive dissonance maintain the educational conditioning of the graduates.

Due to the high number of students, the educational grading system has to be efficient. Therefore, education is organized as a competitive memory-based system. The students are conditioned to compete with each other for praise, status, high marks, and the final prize of a degree. The required knowledge is relative, because their tests and exam results are usually rescaled towards a normal distribution. Being good is not good enough, the student has to be the best in the class. Students are conditioned to win the educational competition, rather than to maximize their individual success. Winning is defeating opponents in competitions. On the other hand, Lao Zi defines success as to obtain what you seek and escape what you suffer [7] (§62). Some people win each competition without ever being successful, while others are successful without ever winning a competition.

In order to win the educational competition, students learn to apply win-lose strategies. Therefore, they rely more on Protagoras' rhetoric than on Socrates' truth [16] [17]. Competition forces students to use rationalizations, rather than Rational Thinking. Rational Thinking aims to find the truth. In contrast, rationalization is counterproductive rhetoric that justifies the thinker's pre-set conclusions in order to win arguments. In addition, competition forces students to replace critical thinking with criticism. Critical thinking is a component of Rational Thinking that predominantly questions the thinker's own descriptions of reality with logic in order to find the truth. On the other hand, criticism is counterproductive rhetoric that aims to reveal true or false weaknesses in others in order to win arguments. Unfortunately, if the eloquent winner is incorrect, then both parties will lose.

The educational system is predominantly memory-based, because the quality of memory can be efficiently evaluated with mass testing. Hence, a good memory fosters educational success. Nevertheless, memory is associated with the past, rather than the present or future. Its value relies on the assumption that events are repeated. Hence, it was very valuable in the repetitive static reality of the past. However, events repeat less often in the modern hyperdynamic reality. Yesterday's solutions are out of date, because today's problems are different. Information has a very short life-span. The moment a student graduates, the degree is already partly outdated.

The test of memory forces students to answer predetermined questions. However, reality does not present prepared questions. In reality, people are expected to find their own questions, before they can even consider an answer. Even worse, providing the right answer for the wrong question is counterproductive. Only after years of intensive conditioning, candidates for post-graduate degrees are allowed to raise one single question. Therefore, the strong educational focus on the efficiency of testing answers inhibits the progress of knowledge.

Thinking is poorly understood. It is difficult to teach and even more difficult to test. Thinking is not even an adequate part of the curriculum. Therefore, high quality thinking does not necessarily maximize a student's educational success. For example, Albert Einstein, who was arguably one of the best thinkers of his era, could not obtain a position in a university after his graduation. Instead, he was forced to accept a job as a second clerk in a Patent Office. To add insult to injury; his teacher Herman Minkowski called him a *'Lazy dog'* [18].

The American mathematician and meteorologist Edward Lorenz (1917-2008) introduced Chaos Theory in 1963. That theory holds that initially minute events can cause huge effects. For example, in the chaotic weather system, the flapping wings of a small butterfly could create a devastating tornado elsewhere [13]. Hence, Lorenz' Butterfly-effect could explain how the huge manmade extinction problems were created by relatively small imperfections in the educational system.

The German philosopher Georg Wilhelm Friedrich Hegel (1770-1831) introduced the dialectic of knowledge [19]. As shown in Figure II: the first stage of his dialectic starts with a thesis, which is a description of reality that is confirmed by observation and/or logic. For example, the Newtonian idea that light is a particle is a thesis [12].

During the second stage of Hegel's dialectic, an antithesis is introduced. An antithesis is a thesis that opposes a preceding thesis. For example, Thomas Young's idea that light is a wave is an antithesis, because it opposes Newton's preceding thesis that light is a particle [20].

Figure II: The development of knowledge

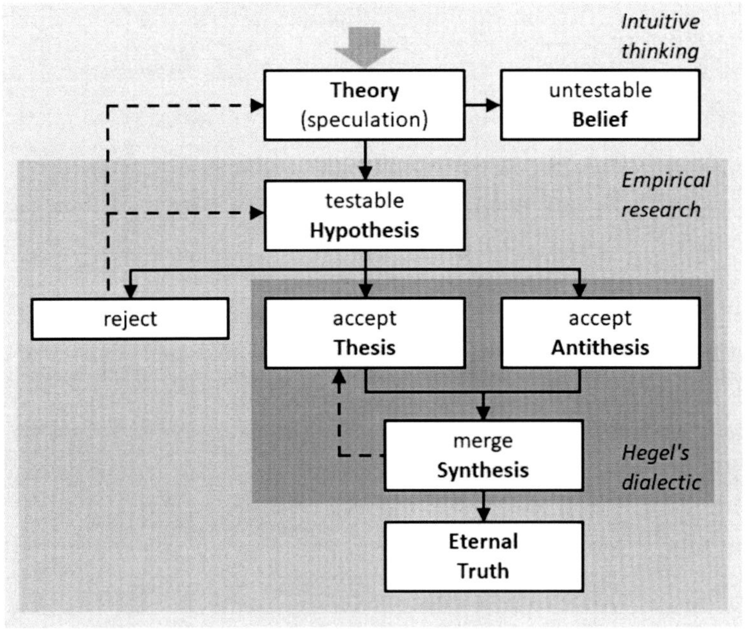

During the third stage of Hegel's dialectic, the conflict between the thesis and antithesis is solved by the introduction of a synthesis. A synthesis is a hypothesis that merges the thesis and antithesis into a new thesis. For example, Albert Einstein's idea that light is both a particle and a wave is a synthesis of Newton's thesis and Young's antithesis [18].

Hegel's dialectic of knowledge will continue until the eternal truth has been reached. The eternal truth is the irrefutable, complete, valid, and reliable description of reality that is tested with observation and/or logic. Unfortunately, thinkers will never know whether their particular thesis is the eternal truth, because they cannot exclude that an antithesis will challenge that thesis in the future. A robust believe in one's *'truth'* fosters the arrogance of static thinking, which inhibit cognitive development. Therefore, Lao Zi warned: *"Arrogance causes misfortune" (§16). Those who are arrogant are without development"* [4] (§24).

Hegel's dialectic is passive, rather than active. It is a static description of the development of knowledge, rather than a dynamic methodology to accelerate that development. As shown in Figure II: the development of knowledge starts with a theory, rather than with a Hegelian thesis. A theory is a speculative description of reality that leads to either a belief or a hypothesis. A belief is a description of reality that cannot be tested with observation or logic. In contrast, a hypothesis is an untested description of reality that can be tested with observation and/or logic.

Cognitive Accelerator

Ideas of Lao Zi, Aristotle, and Hegel are synthesized in order to create the Cognitive Accelerator. That accelerator fosters the cognitive revolution by turning counterproductive static thinking into productive dynamic thinking.

- During the first stage—dynamic thinkers formulate a hypothesis with Affectorial Thinking. Subsequently, they subject that hypothesis to Rational Thinking in order to create a thesis. Ultimately, the development of knowledge cannot start with Rational Thinking.
- During the second stage—dynamic thinkers challenge the thesis by searching for the antithesis. They might play the Devil's advocate. In contrast, competition drives static thinkers to win arguments. Therefore, they are prone to stagnate in a counterproductive defence of the thesis. For example, defending Newton's particle theory of light stagnated the development of knowledge for centuries.
- During the third stage—dynamic thinkers search for the synthesis that merges the thesis with the antithesis. Therefore, they focus on similarities. The synthesis becomes automatically the new thesis, which continues the acceleration of knowledge. In contrast, competition drives static thinkers to counterproductive arguments about the differences between thesis and antithesis. For example, the argument about the particle and wave theories of light stagnated the development of knowledge for decades.

In order to maximize the effect of the Cognitive Accelerator: students should be reinforced to create productive questions. Education should teach them to maximize their success, rather than to win competitions. Students should be taught to search for the truth, rather than to take semantic and rhetorical short-cuts. In accord, teachers should help them to replace counterproductive rationalizations with Rational Thinking, and criticisms with critical thinking. Teachers should motivate students to be confident and rely on their creativity, intuition, emotions, and skills provided by their Affectorial Thinking.

Dynamic thinking is a precious asset in the modern hyperdynamic reality. Therefore, it should become a formal part of the educational curriculum. A proper test of the student's ability to apply dynamic thinking should become the basis for the allocation of socio-economic positions. The educational system should use conditioning to install groupthink and cognitive dissonance in order to spread and maintain dynamic thinking.

Summary

The technological revolutions have transformed the static reality of the past into the modern hyperdynamic reality. Humanity's way of thinking is failing, because they have become static thinkers in a hyperdynamic reality. That systemic cognitive failure creates a growing host of manmade extinction problems such as nuclear war and climate change. Those problems cannot be solved with the same way of thinking that has created them. Lao Zi warned that knowledge is a two-edged sword: "*The more people know, the stranger the things they begin to develop*" [1, p. §57].

Unfortunately, educational conditioning, dogmatism, groupthink, and cognitive dissonance inhibit the transformation from static to dynamic thinking. This paper introduced the Cognitive Accelerator in order to initiate a cognitive revolution that supports the next generation in their Herculin task to solve the manmade extinction problems.

References

[1] Aristotle, Organon, Complete Edition (Compiled by the Peripatetics), Createspace Independent Publishing Platform, 2015.

[2] Schade, A., Think Smarter with Nemonik Thinking, Dunedin: nemonik-thinking.org (Printer Amazon), 2016b.

[3] Schade, A., Dictionary Lao Zi's Dao De Jing, 1 ed., vol. 1, Dunedin: nemonik-thinking.org (Printer Amazon), 2018a.

[4] Schade, A., Meta-translation Lao Zi's Dao De Jing (1-37), 3 ed., vol. 2, Dunedin: nemonik-thinking.org (Printer Amazon), 2018b.

[5] Schade, A., Meta-translation Lao Zi's Dao De Jing (38-81), 3 ed., vol. 3, Dunedin: nemonik-thinking.org (Printer Amazon), 2018c.

[6] Schade, A., Lao Zi's Dao De Jing Demystified, 2 ed., vol. 4, Dunedin: nemonik-thinking.org (Printer Amazon), 2017a.

[7] Schade, A., Stunning Revelations about Lao Zi's Dao De Jing, 2 ed., vol. 5, Dunedin: nemonik-thinking.org (Printer Amazon), 2017b.

[8] Zhuang Zi, The Complete Works of Chuang Tzu, New York: Columbia University Press, 1968.

[9] Zhuang Zi, Chuang Tzu, London: Unwin Paperbacks, 1980.

[10] Skinner, B. F., About Behaviorism, New York: Knopf (distributor Random House), 1974.

[11] Wallace, W. A., Domingo de Soto and the Early Galileo, Aldershot: Ashgate Publishing, 2004.

[12] Newton, Sir Isaac, The Mathematical Principles of Natural Philosophy (Principia), London, 1687.

[13] Lorenz, E. N., "Deterministic nonperiodic flow," Journal of the Atmospheric Sciences, vol. 20, p. 130–141, 1963.

[14] von Goethe, J. W., Der Zauberlehrling (The Sorcerer's Apprentice), Poem, 1797.

[15] Schade, A., Education Kills Humanity, Dunedin: nemonik-thinking.org (Printer Amazon), 2016a.

[16] Koyre, A., Discovering Plato, New York: Columbia University Press, 1945.

[17] De Crescenzo, L., The history of Greek Philosophy, London: Arnaldo Mondadori Editore S. p. A., 1983.

[18] Gribben, J., Get a Grip on New Physics, London, England.: Weidenfeld and Nicolson, 1999.

[19] Chalybäus, H. M., Historische Entwicklung der spekulativen Philosophie von Kant bis Hegel, Edinburgh: T. & T. Clark, 1854.

[20] Young, T., "The Bakerian Lecture. Experiments and Calculations relative to physical Optics," Philosophical transactions, 1809.

Endnotes

1 Logic is a cognitive method that consists of logical arguments. Each argument comprises a set of facts that result in an inevitable valid and reliable conclusion. Each conclusion becomes a new fact and so on. For example (A > B, B > C, therefore A > C). A fact is a valid and reliable description of reality that is tested with logic. Valid means that the description of reality describes what is supposed to describe. Reliable means that the description of reality is accurate and consistent across time.

2 老子 (Lao Zi or Lao Tzu). 道德經 (Dao De Jing or Tao Te Ching). Dao means: Path, Way, Way of Nature, or Physics. De means: Virtue, Way of People, or Psychology. Jing means: Classic. Hence, *Dao De Jing* means: *A Classic about the Physics of Psychology.*

3 古 (the Ancients).

4 吾言什易知也 什易行也 而人莫之能知也 而莫之能行也 (§70).

5 Six Chinese versions were analysed and integrated: (Wang Bi, 226-249 AD); (He-Shang Gong, 179-157 BC); (Fu Yi, 555-639 AD); (Mawangdui-A, ~200 BC); (Mawangdui-B, ~200 BC); and (Guodian, ~300 BC).

6 Nine English translations were analysed and integrated: (Chan Wing-Tsit, 1988); (Cheng Gia-Fu and English, J, 1972); (Henricks, 1993); (Land, 1990); (Lau, 1985); (Lin, J. P., 1977); (Man-ho Kwok; Palmer, M.; & Ramsay, J., 1997); (Waley, 1968); and (Wing, 1986).

7 For example, Lao Zi refers to *Gravity* as the *Mysterious Female.*

8 孔德之容唯道是从 (§21).

9 毋遗身央是谓袭常 (§52). The '*constant* (常)' is a synonym for Dao, the Laws of nature, or Physics.

10 不道早已 ((§30 & 55).

11 After the fall of the Latin speaking Western Roman Empire during the 5th century, most of Aristotle's work was lost in the West. His work was rediscovered during the Renaissance period that lasted from the 14th to the 17th century.

12 求以得有罪以免与 (§62).

APPENDICES

My Journey

Tales about my youth might help you on your journey. I was born in Amsterdam at the beginning of World War II. It was a kind of false start. It was the wrong time at the wrong place. Lucky for me, I had the right parents. They were true survivors and kept me alive during the hardship. Lao Zi warned: *"Do not use soldiers to force the world. Such actions are likely to rebound. Where armies have camped only thorny bushes will grow"* [1] *§30.* Unfortunately, humanity had not learned after two-and-halve thousand years. Therefore, the end of that war became the *'Hunger-Winter'.*

Food was almost non-existent. If you had some, the German army confiscated it without any grace. Those soldiers were also hungry. Instead of fighting on the front, they were hunting rabbits in the fields around Amsterdam. It was a time that all the cats disappeared from the neighbourhood. They were called *'roof-rabbits'* and went in a steady procession to rabbit heaven. It was not a good time for children with pets. Lucky for me, I was poor and had none.

We were fortunate to have overcooked tulip-bulbs for breakfast, lunch, and dinner. My father, Sijbrand, was an industrious man. Most probably, he dug them up in the nearby Vondel Park. You remember the song: *Tulips from Amsterdam.* It has a different meaning for me. It reminds me of their strong sweet smell filling our noses and the entire house. That smell was like glue and stuck to everything. But beggars can't be choosers. Even sweet smelly bulbs were much better than the alternative of death by starvation.

One day, my father came home from his forage with boiled potatoes wrapped in and old newspaper. It was truly unbelievable. I never tasted a better meal. There was enough for everyone. It was like winning the Lotto on Christmas day. He must have been starving like all of us, but strangely enough, he did not eat anything. He just sat there at the head of our small table and encouraged us to eat. Years after the war, we prodded him about that day. He laughingly confessed that he had retrieved the potatoes from a dirty rubbish bin and could not bring himself to eat them. Without using any words, my father taught me to be practical.

The hunger-winter was also extremely cold and there was not enough fuel left in Amsterdam to heat the houses. Most fuel was confiscated by the German forces. At that time, the counterattack of their panzer divisions in the Ardennes was stopped by a lack of fuel. At least, the shortage did some good.

In my neighbourhood, all trees, benches, fences, and garden sheds disappeared in the heaters to fight the remorseless cold. Even the wooden supports between the tram rails disappeared. Worse, the toys of my seven years older brother were also fed into the heater. Imagine his agony. There was no choice. I was lucky again. I had none. It taught me that in a hostile environment you cannot remain a child.

I survived World War II by the skin of my teeth and went to school during the beginning of the Cold War. Probably, my teachers had never seen the pictures of Hiroshima and Nagasaki after the nuclear explosions. They were either optimists or laughing idiots because they told us that the Russians were coming. Therefore, we had to hide under our flimsy desks from ***'The Bomb'***. Those exercises scared the living daylights out of me. Hopefully, the false feeling of security made my teachers feel better. After that experience, life took a wrong turn. Nothing that good can last forever.

In his early sixties, my father stepped on a cheap soapbox on a busy footpath in Amsterdam and addressed the passers-by. He advised them to maintain the hard-won peace and warned them to take care of nature. In his shabby raincoat, he was the prototype Pacifist and Greenie all rolled in one.

I watched him addressing the public when the police arrived. Dressed in their fashionable and spotless black uniforms, they were short tempered and did not argue or explain. Although my father did not resist, they took out their truncheons and beat him up. I felt every blow as if it hit me.

In the end, they threw him in the back of their black van like a common criminal. To them, it was just light exercise—another routine case. For me, it turned the world upside down. I was stunned. Yesterday, the Germans soldiers were the enemy. Today, the Dutch police should protect people like my father. Right? But what did I know? I was just a powerless child that did not understand politics.

Nevertheless, that beating was only the beginning. Like in Stalin's Russia, the authorities imprisoned my father in an asylum where he received electric shock treatment for his *'dangerous ideas'*. It was nothing less than torture under the thin cloak of medical treatment. I did not see him for the next six months.

Being a true stubborn Frisian, my father did not give up. He repeated the process five times. The authorities were evenly stubborn and *'treated'* him also five times in return. At least, they did not put him on the cross. My mother, Henderika, was a wise woman. She taught me to be grateful for such small blessings.

Those atrocities happened in the outskirts of peaceful Amsterdam, rather than in that cruel Nazi Berlin that we fought with blood and tears in order to regain our freedom. My father's fate taught me that every political system, from democracy to tyranny, defends its narrative and groupthink ultimately with brutal force.

Lao Zi warnings are still true today: *"Superior propriety acts and if there is no agreement, then the arms are bared"* [1] *§38*. *"When people do not fear authority, then greater authority will appear"* [1] *§72*. I deeply respect my father for using his mental strength to stand up for the causes he believed in. Fortunately, those authorities failed, because seventy years later, I am still standing.

My father must have affected my thinking a lot. I remember, letting the kitchen tap run in order to return clean drinking water back to nature. My idea was to reduce the manmade water pollution. That must have cost him a fortune, but then teachings of high morality and ethics do never come cheap.

My father was a great role-model in many ways. He showed me that, even in an emergency, waiting can be far more productive than action. People often act rashly, because they are weak and cannot cope with the stress of waiting. Fear is the killer of the little mind. People who are biased towards action, do not seem to know that each action creates a counterproductive reaction. They jump from the pan straight into the fire.

It was a few years after the war. We lived next to one of Amsterdam's many concentric canals. It was a quiet time. There were no tourists yet. Farmers still transported their produce with small boats with puffing outboard motors through the canal in order to supply the market in the centre of Amsterdam.

During one of the long and cold winters, the water was frozen solid. Gigantic icebreakers, belching hot steam and black smoke from their huge chimneys, thundered to the rescue of Amsterdam's lifeline. Under full steam with roaring engines, they rammed their huge steel bows on the thick ice again and again. The enormous weight of those juggernauts split the solid ice sheet with thunderous noise into floats.

For us kids, it was a true spectacle. It was like watching the Titanic running in an iceberg in our own backyard. Much better than the circus. This was real and free—cracking ice and violent waves moving in all directions. The quayside was packed with a crowd of onlookers and school could wait. Sometimes, we could not get fast enough out of the way and the waves drenched out feet.

The next day, my friend and I decided that it was fun to cross the canal by jumping from one ice-float to another. Imagine that. You can't just walk to the edge of a float. Your weight will flip it upside down. You have to jump from centre to centre. Total madness! Slippery ice-floats moving in all directions with two underfed street urchins jumping from one float to the next.

It was like maintaining your balance on the point of a needle. One tiny slip and the thick slabs of ice would have closed over our heads like a steel trap until next summer. My hands get sweaty if I think about it. Just for fun? What were we thinking? We had a lot to learn. Nearby, was a proper bridge across the canal.

We were already a few meters on our way to a certain death when my father turned up. He just stood there at the quayside and signalled us by crooking his finger to come back to him. As he acted totally cool, calm, and collected, we just move towards him. Easy as pie. He bent over, grabbed our arms, and pulled us both of the ice. Without even raising his voice, he told us never to do that again and sent us home. As far as I know, he never worried my mother with our misadventure. All in a day's work.

If he would have jumped on the ice, then we all would have slipped under it and died. If he would have shouted and showed his anxiety, then my friend and I would have panicked with the exact same result. Death! It would have been over in seconds. He managed his fear and did not take any action. He just stood there and waited.

What must have gone on in his head? Poor guy! Imagine that one of your loved ones was balancing on those floats. My father stepped up to the proverbial plate and was a true hero. As a result, all three of us survived. That taught me that action without great action or bravado can save any day.

"Act with Non-action. Work without effort" §63. "Those who act will fail. Those who hold will lose. Therefore, sages will use Non-action. Therefore, they will not fail" [1] *§64.*

We live in a reality that consist of interacting and precarious balances that act like chains of dominos. In that reality, each action evokes a counterproductive reaction in order to restore the balance. Therefore, Lao Zi wrote: *"...the only thing I fear is action"* [1] *§53.* Unfortunately, this is an era that glorifies action. As a result, imbalances threaten directly our survival.

When I was about twelve, my family still suffered from the extreme post-war poverty. Therefore, I was far more interested in food than education. Our meagre diet was just one tiny step above the notorious tulip-bulbs. My body was desperately trying to grow, but instead was hungry all the time.

At school I did not pay any attention to the lessons in music and proper language. As punishment, I am tone-deaf and grammar is still a headache. I was just waiting to get out. My teacher must have been a genius, because he wrote in my annual report: *"Auke is watching the birds"*. How did he know? By the way, that were real birds, rather than girls.

One day, I was stupid enough to outsmart my math teacher. In revenge, he accused me of cheating. Why would I cheat? I was not even interested in his teachings. In addition, I never had a need to cheat. I was a lucky child, my mind always solved automatically what had to be solved in order to survive. He should have searched for the child that I was supposed to have cheated from. Lao Zi wrote: *"Those who do not value their teachers, or do not love their students, although knowledgeable, are greatly confused" §27. "Those who know are not educated. Those who are educated do not know"* [1] *§80.* The utter confusion of my knowledgeable teacher taught me to question every judgement.

To earn some pocket money, I helped the neighbourhood cobbler on my free Saturdays returning the repaired shoes to his clients. In addition, during the Christmas period, I roamed around the local saleyards of Christmas trees and offered the buyers to carry their trees home. However, I had to stay away from the sellers, because they saw me as a nuisance disturbing their trade. Given halve a chance, they would clip me around the ears.

Another time, I helped the grumpy shop owner next door to clean up his grocery store. After I was finished, he grudgingly gave me a chocolate bar. That means little today, but at that time it was a huge and expensive treat. I had never tasted chocolate before. I raced home and proudly opened it together with my mother. Inside, was a block of wood. It was a fake window display.

My mother was furious and run off. I have never seen her angry, but that one time she was fuming. I do not know what she did to the shop keeper. I still fear the worse for him. If my mother would not get him, then the Devil would. Honestly, I think that he would have preferred the Devil. My mother was an angel with a sharp sword. I learned from his false generosity that you have to set very clear conditions before you agree working for anyone. Lao Zi warned: *"Those with virtue will uphold the agreement. Those without virtue will uphold the details"* [1] *§79.*

At age thirteen, I had to walk to my technical school to study carpentry. Four times a day for about half an hour. That winter, the footpath was covered with about halve a meter of snow and ice, while I was dressed in my only thin summer shorts and shoes with huge holes in them. Actually, they were more like holes with the remnants of shoes hanging on around the edges. I must have been an optimist, because I put carton on the inside to stop the water and cold getting in. Fortunately, I could not get hyperthermia, because that concept was not discovered yet.

When I woke up in the morning, the windows in my bedroom were covered with a thick layer of ice-flowers. They were a beautiful and lethal threat. In my recurrent nightmares during that period, I was stuck in a dark cave and the only exit was surprisingly locked with ice. Lucky enough, I had a small nail and tried every night to scratch my way through the solid ice. I have to admit, I never succeeded. In another reality, I might be still hoping to get out. Hence, the characters described by Charles Dickens had nothing on me. I learned the hard way what hunger and cold really means. How it saps relentlessly your strength and determination. That taught me Lao Zi's three treasures.

"I have always three treasures that I keep and protect. The first one is called compassion. The second one is called frugality. The third one is called humbleness" [1] *§67.*

At sixteen, I worked during the day as a carpenter and studied engineering at night. I received my first decent meal when I was nineteen years old, because I was drafted into the army. It was a large plate filled to the brim with steaming white rice, green vegetables, and juicy bits of chicken. Despite Lao Zi's warnings against armies, it was Heaven on Earth. Hallelujah! That moment, my life turned for the better. The price for all that luxury was that I had to interrupt my engineering study for two years. During the seventies, I was a financial manager for an international company and set up my own retail business. In my free time, I studied economy.

In 1980, I resigned and left everything behind. I immigrated with my family from the Netherlands to New Zealand in order to escape the nuclear threat in Europe. Fortunately, the threat never became reality, but we escaped the nuclear fallout from Chernobyl.

In my new country, I started again and went to university to study psychology. This time my teachers did not accuse me of cheating. Instead, they appreciated my work and I never had to finish my master's degree. Halfway that study, the university transferred my Master's into a PhD. I must have done something right.

Life has been a roller coaster. I have been high and low. I have seen most of it if not all. The start of my life might have been somewhat tough, but I would not change anything, because that pressure made me what I am right now. That is why I have a story to tell.

I have learned from painful experiences that I can only maximize my success if I manage my mind. Therefore, I have compiled my survival strategies in Nemonik Thinking [5] [6]. Nemonik thinking is a conscious, semiconscious, and subconscious way of thinking that uses Dynamic Thinking and Mind Management to maximize external and internal success. It worked for me. Nemonik Thinking turned the chaos around me into order. Despite the pressure, I have never been depressed. In addition, I have been free of smoking, alcohol, and drugs for more than eighty years. I never had a need for those escapes. You could save yourself time and agony by learning from my mistakes. Alternatively, you could make your own. The choice is yours.

Abstracts of my other Books

Glossary Nemonik Thinking [5]

We need clear definitions to communicate. However, definitions associated with the mind and reality are inherently hypothetical, fuzzy, and intertwined. Therefore, this glossary attempts to provide definitions of the main concepts associated with nemonik thinking. Nemonik thinking will improve your life. It accelerates your thinking; improves your memory; mobilizes your subconscious genius; strengthens your weaknesses; and reveals opportunities and threats. It helps you to pursue your goals in the right state of mind; at the right place, at the right time, with the right resources; and the right information. Nemonik thinking will assist you to think on your feet and become panic resistant during emergencies. Nemonik thinking is simple to learn, but it so sophisticated that it will make you a superior problem solver. Nemonik thinkers evaluate a checklist of seventeen (now 21) nemoniks for each situation. Nemoniks are memorized keywords describing all aspects of your mind, reality, and their interaction (Schade, Think Smarter with Nemonik Thinking, 2016) [6].

Free eBook @

nemonik-thinking.org

Think Smarter with Nemonik Thinking [6]

Nemonik thinking will improve your life. It accelerates your thinking; improves your memory; mobilizes your subconscious genius; strengthens your weaknesses; and reveals opportunities and threats. It helps you to pursue your goals in the right state of mind; at the right place, at the right time, with the right resources; and the right information. Nemonik thinking will assist you to think on your feet and become panic resistant during emergencies. Nemonik thinking is simple to learn, but it so sophisticated that it will make you a superior problem solver. Nemonik thinkers evaluate a checklist of seventeen (now 21) nemoniks for each situation. Nemoniks are memorized keywords describing all aspects of your mind, reality, and their interaction. Nemonik thinking is like playing a musical keyboard with seventeen (now 21) keys producing an infinite repertoire of practical strategies. Nemonik thinking provides the strategic options for any possible situation, while Lao Zi's Dao De Jing provides the principle that identifies which of those options will fit the actual situation (Schade, Stunning Revelations about Lao Zi's Dao De Jing, 2017) [7]. The resulting strategies will maximize your success, which is to obtain what you seek and escape what you suffer. You might be the best thinker in the world, but only nemonik thinking could make you the smartest thinker you can be.

Free eBook @

nemonik-thinking.org

Dictionary Lao Zi's Dao De Jing [13]

This Chinese-English and English-Chinese dictionary is especially compiled for the translation of Lao Zi's ancient book Dao De Jing (Schade, Meta-translation Lao Zi's Dao De Jing (1-37), 2018 [14]) and (Schade, Meta-translation Lao Zi's Dao De Jing (38-81), 2018 [15]). Dao De Jing means literally—A Classic about the Way of Nature and the Way of People. It aims to maximize your success, which is to obtain what you seek and escape what you suffer. Success is maximized by aligning the Way of People with the Way of Nature (Schade, Stunning Revelations about Lao Zi's Dao De Jing, 2017 [7]). The Chinese versions used in this study include—(Fu Yi, ~200 BC); (Guodian, ~300 BC); (He-Shang Gong, 179-157 BC); (Mawangdui (A), ~200 BC); (Mawangdui (B), ~200 BC); and (Wang Bi, 226-249 AD). Together, those versions contain about 1,600 different pictographs. Every language changes over time and, therefore, some of Lao Zi's ancient pictographs are not used anymore, while the meaning of others has changed. In addition, most modern Chinese pictographs have several English meanings that foster ambiguity. Therefore, the exhaustive English meanings for each pictograph were extracted from reputable sources. Furthermore, a unique system of Digital Index for Pictographs (DIP) is introduced that simplifies the digital classification of Chinese pictographs.

Free eBook @

nemonik-thinking.org

Meta-translation Lao Zi 1-37 [14]

The title of Lao Zi's ancient book Dao De Jing means literally—A Classic about the Way of Nature and the Way of People. Dao De Jing aims to maximize your success, which is to obtain what you seek and escape what you suffer. Success is maximized by aligning the Way of People with the Way of Nature (Schade, Stunning Revelations about Lao Zi's Dao De Jing, 2017 [7]). Despite the great efforts, previous translations of Dao De Jing do not present an adequate understanding of that mysterious manuscript. In order to take optimal advantage of the expertise accumulated in such earlier studies, this meta-translation is based on an English meta-analysis and a Chinese meta-analysis. The English meta-analysis is based on the following English translations—(Chan Wing-Tsit, 1988); (Cheng Gia-Fu and English, J, 1972); (Henricks, Lao-Tzu: Te-Tao Ching, 1993); (Land, 1990); (Lau, Lao Tzu: Tao Te Ching, 1985); (Lin, J. P., 1977); (Man-ho Kwok; Palmer, M.; & Ramsay, J., 1997); (Waley, 1968); and (Wing, 1986). The Chinese meta-analysis is based on the following Chinese versions of Dao De Jing—(Fu Yi, ~200 BC); (Guodian, ~300 BC); (He-Shang Gong, 179-157 BC); (Mawangdui (A), ~200 BC); (Mawangdui (B), ~200 BC); and (Wang Bi, 226-249 AD). This meta-translation of Dao (Chapters 1-37) is based on a special dictionary (Schade, Dictionary Lao Zi's Dao De Jing, 2018 [13]), while it is the foundation for (Schade, Lao Zi's Dao De Jing Demystified, 2017 [15]).

Free eBook @

nemonik-thinking.org

Meta-translation Lao Zi 38-81 [15]

The title of Lao Zi's ancient book Dao De Jing means literally—A Classic about the Way of Nature and the Way of People. Dao De Jing aims to maximize your success, which is to obtain what you seek and escape what you suffer. Success is maximized by aligning the Way of People with the Way of Nature (Schade, Stunning Revelations about Lao Zi's Dao De Jing, 2017 [7]). Despite the great efforts, previous translations of Dao De Jing do not present an adequate understanding of that mysterious manuscript. In order to take optimal advantage of the expertise accumulated in such earlier studies, this meta-translation is based on an English meta-analysis and a Chinese meta-analysis. The English meta-analysis is based on the following English translations—(Chan Wing-Tsit, 1988); (Cheng Gia-Fu and English, J, 1972); (Henricks, Lao-Tzu: Te-Tao Ching, 1993); (Land, 1990); (Lau, Lao Tzu: Tao Te Ching, 1985); (Lin, J. P., 1977); (Man-ho Kwok; Palmer, M.; & Ramsay, J., 1997); (Waley, 1968); and (Wing, 1986). The Chinese meta-analysis is based on the following Chinese versions of Dao De Jing—(Fu Yi, ~200 BC); (Guodian, ~300 BC); (He-Shang Gong, 179-157 BC); (Mawangdui (A), ~200 BC); (Mawangdui (B), ~200 BC); and (Wang Bi, 226-249 AD). This meta-translation of De (Chapters 38-81) is based on a special dictionary (Schade, Dictionary Lao Zi's Dao De Jing, 2018 [13]), while it is the foundation for (Schade, Lao Zi's Dao De Jing Demystified, 2017 [16]).

Free eBook @

nemonik-thinking.org

Lao Zi's Dao De Jing Demystified [16]

The title of Lao Zi's ancient book Dao De Jing means literally—A Classic about the Way of Nature and the Way of People. It aims to maximize your success, which is to obtain what you seek and escape what you suffer. Success is maximized by aligning the Way of People with the Way of Nature. The current study presents four English versions of increasing demystification. The first one is the Meta-translation version as developed in (Schade, Meta-translation Lao Zi's Dao De Jing (1-37), 2018 [14]) and (Schade, Meta-translation Lao Zi's Dao De Jing (38-81), 2018 [15]). However, Lao Zi's poetic style, mysticism, metaphors, and synonyms are still inhibiting a clear understanding. The second one is the Clarification version, which is presented in parallel with the Meta-translation. That clarification increased the consistency of Lao Zi's terminology, but significant chapters about Dao are still located in the De section and vice versa. Therefore, the third one is the Chapters Reorganized version in which the chapters are relocated to the Dao and De sections. Nevertheless, the text remains fuzzy, because several chapters relate to both sections. Therefore, the fourth one is the Sentences Reorganized version in which the sentences are sorted by topic. That version is the foundation for (Schade, Stunning Revelations about Lao Zi's Dao De Jing, 2017 [7]).

Free eBook @

nemonik-thinking.org

Stunning Revelations about Lao Zi [7]

The title of Lao Zi's ancient book Dao De Jing means in modern terminology—A Classic about the Physics of Psychology. Dao De Jing is a significant contribution of the Chinese literature to the contemporary sciences of physics and psychology. Lao Zi presents a sophisticated theory concerning the origin, formation, and working of the universe. Although that theory is at the cutting edge of modern physics, it provides a simple and practical principle. That principle holds that the unstoppable force Qi will always maintain the multitude of Yin-Yang balances comprising our universe. Therefore, Lao Zi's theory of psychology predicts that people can only maximize their success by aligning with Qi. Success is to obtain what you seek and escape what you suffer. Unfortunately, humanity has ignored Lao Zi's principles for two-and-halve thousand years. As a result, we are now facing manmade problems such as climate change, continuous warfare, dwindling resources, environmental pollution, and overpopulation. Those problems threaten your personal success and they cannot be solved with the same way of thinking that has created them. Therefore, Lao Zi's Dao De Jing is more relevant than ever. His way of dynamic thinking fosters solutions that reach peacefully across the fault-lines created by race, religion, and ideology. This book is based on (Schade, Lao Zi's Dao De Jing Demystified, 2017 [16]).

Free eBook @

nemonik-thinking.org

Lao Zi's Dao De Jing (Chinese-English) [17]

Lao Zi's Dao De Jing or Lao Tzu's Tao Te Ching is a beautiful example of ancient Chinese literature. Dao De Jing means literally—A Classic about the Way of Nature and the Way of People. It aims to maximize your success, which is to obtain what you seek and escape what you suffer. Success is maximized by aligning the Way of People with the Way of Nature. After two-and-halve thousand years, Lao Zi's wisdom is still ahead of time and outshines intellectual giants such as Confucius, Sun Zi, Socrates, Plato, and Aristotle. Lao Zi's deep understanding of nature and people is your guiding light to a better future, because his way of dynamic thinking reaches peacefully across the fault-lines created by race, religion, and ideology. Therefore, Dao De Jing is more relevant than ever. This book comprises Chinese and English versions of Dao De Jing, which are extracted from (Schade, Lao Zi's Dao De Jing Demystified, 2017 [16]). Furthermore, Lao Zi's amazing secrets are revealed in (Schade, Stunning Revelations about Lao Zi's Dao De Jing, 2017 [7]).

Free eBook @

nemonik-thinking.org

Lao Zi's Poems [18]

The Chinese philosopher Lao Zi wrote two-and-halve thousand years ago the brilliant book Dao De Jing, which means literally—A Classic about the Way of Nature and the Way of People. That ancient book contains eighty-one poems of extraordinary depth describing a natural way of living that provides long-term success with minimum effort—*The Way of Nature has no favourites. It is always with the competent people* (§79). The theme of his poems is to maximize your success, which is to *obtain what you seek and escape what you suffer* (§62). You maximize your success by aligning the Way of People with the Way of Nature—*Have a long life through a lasting regard for the Way* (§59). Manmade climate change is only one of many consequences of ignoring Lao Zi's wisdom. His superb poems form a bridge across the deep fault lines that are created and maintained by nationality, ideology, religion, race, and greed. The present book comprises an extract from (Schade, Lao Zi's Dao De Jing Demystified, 2017 [16]).

Free eBook @

nemonik-thinking.org

Education Kills Humanity [2]

The aim of this study is to evaluate the educational system with nemonik thinking (Schade, Think Smarter with Nemonik Thinking, 2016 [6]). The system conditions students with certificates to maximize their probability of winning the educational competition. The winners are rewarded with advantageous positions in society, while the losers are doomed to serve the winners. Therefore, conventional thinking is conflict oriented, which fosters aggression, control, effort, and force. The conditioned compulsion to win arguments inhibits also the truth and, therefore, it corrupts conventional thinking. As a result, the educational system produces incompetent leaders and rebellious followers. That counterproductive combination fosters manmade problems such as climate change, continuous warfare, dwindling resources, environmental pollution, and overpopulation. Those problems cannot be solved with the same way of conventional thinking that has created them. In contrast, nemonik thinking aims for success, rather than winning. Success is to obtain what you seek and to escape what you suffer. Therefore, nemonik thinking is goal oriented, which fosters freedom, alignment, compassion, allies, and win-win strategies. You might be the best thinker in the world, but only nemonik thinking could make you the smartest thinker you can be. This book contains extracts from (Schade, Think Smarter with Nemonik Thinking, 2016 [6]).

Free eBook @

nemonik-thinking.org

Global Warming is the Solution [3]

The aim of this study is to accelerate the development of climatology with nemonik thinking (Schade, Think Smarter with Nemonik Thinking, 2016 [6]). About 400,000 years of data, extracted from the Antarctic Vostok ice-core, were subjected to statistical analyses. The results suggest that the duration and thermal stability of the current interglacial are significantly larger than those of the four previous ones are. Hence, the results could not be contributed to natural variables. This supports the hypothesis that the current interglacial period changed some time ago into a glacial period, while artificial global warming has compensated for that natural global cooling. During a glacial period, the average global atmospheric temperature could decrease with about 10.0 °C. This would reduce the global food supply; threaten the global infrastructure; and force billions of people to migrate back towards the equator. Therefore, artificial global warming could be the solution for natural glacial cooling. Further research of this topic is crucial.

The Threat of Bilateral Climate Change [4]

The aim of this study is to accelerate the development of climatology with nemonik thinking (Schade, Think Smarter with Nemonik Thinking, 2016 [6]). During the industrial period, the atmospheric CO2 concentration has increased about 120 ppm , while the average global atmospheric temperature increased about 1.4 °C . However, the Vostok thermal function of CO2 during the last 400,000 years predicts that the increase of 120 ppm of CO2 would increase the temperature with 11.6 °C . Hence, there is a thermal gap of 10.2 °C between the observed and predicted temperature. This gap could be explained by the proposed bilateral hypothesis of climate change. This hypothesis holds that the observed increase in average global atmospheric temperature of 1.4 °C is the balance of an artificial global warming of 11.6 °C and a natural global cooling of 10.2 °C . Those large opposing thermal phenomena could explain the recent climatological instability. Furthermore, the bilateral hypothesis predicts that an uncontrolled decrease of atmospheric CO2 could trigger glacial conditions threatening humanity. Therefore, further research of this topic is crucial. This book contains extracts of (Schade, Global Warming is the Solution, 2016 [3]).

Free eBook @

nemonik-thinking.org

References

[1] Schade, A., Lao Zi's Dao De Jing (In: Maximum Success with Nemonik Thinking - Ancient Wisdom), Dunedin NZ: nemonik-thinking.org, 2022.

[2] Schade, A., Education Kills Humanity, Dunedin NZ: nemonik-thinking.org (Printer Amazon), 2016.

[3] Schade, A., Global Warming is the Solution, Dunedin NZ: nemonik-thinking.org (Printer Amazon)., 2016.

[4] Schade, A., The Threat of Bilateral Climate Change, Dunedin NZ: nemonik-thinking.org (Printer Amazom), 2016.

[5] Schade, A., Glossary Nemonik Thinking, Dunedin NZ: nemonik-thinking.org (Printer Amazon), 2016.

[6] Schade, A., Think Smarter with Nemonik Thinking, Dunedin NZ: nemonik-thinking.org (Printer Amazon), 2016.

[7] Schade, A., Stunning Revalations about Lao Zi's Dao De Jing, 2 ed., vol. 5, Dunedin NZ: nemonik-thinking.org (Printer Amazon), 2017.

[8] Schade, A., "Ancient Theory Challenges Modern Physics (In: Maximum Success with Nemonik Thinking - Ancient Wisdom)," *Unpublished Paper,* 2021.

[9] Schade, A., "Ancient Theory Challenges Modern Psychology (In: Maximum Success with Nemonik Thinking - Ancient Wisdom)," *Unpublished Paper,* 2021.

[10] Schade, A., "The Cognitive Revolution (In: Maximum Success with Nemonik Thinking - Ancient Wisdom)," *Unpublished Paper,* 2021.

[11] Schade, A., Nemonik Information System (NIS), Dunedin NZ: nemonik-thinking.org, 2021.

[12] Schade, A., Nemonik Evaluation System (NES), Dunedin NZ: nemonik-thinking.org, 2018.

[13] Schade, A., Dictionary Lao Zi's Dao De Jing, Dunedin NZ: nemonik-thinking.org (Printer Amazon), 2018.

[14] Schade, A., Meta-translation Lao Zi's Dao De Jing (1-37), 3 ed., vol. 2, Dunedin NZ: nemonik-thinking.org (Printer Amazon), 2018.

[15] Schade, A., Meta-translation Lao Zi's Dao De Jing (38-81), 3 ed., vol. 3, Dunedin NZ: nemonik-thinking.org (Printer Amazon), 2018.

[16] Schade, A., Lao Zi's Dao De Jing Demystified, 2 ed., vol. 4, Dunedin NZ: nemonik-thinking.org (Printer Amazon), 2017, p. 478.

[17] Schade, A., Lao Zi's Dao De Jing (Chinese-English), Dunedin NZ: nemonik-thinking.org (Printer Amazon), 2016.

[18] Schade, A., Lao Zi's Poems, Dunedin NZ: nemonik-thinking.org (Printer Amazon), 2018.

Glossary

ACCEPT—nemonik that prompts the mind to maximize success by accepting the incoming information as a true description of the reality.

ACCUMULATE—nemonik that prompts the mind to maximize success by increasing the amount of matter that is under control.

ACT—nemonik that prompts the mind to maximize success by changing or moving matter in space and time.

ADVANCE—nemonik that prompts the mind to maximize success by decreasing the distance to the goal.

Affecter—mental signal generated by a subconscious mindset that influences the conscious.

Affectorial Thinking—subconscious way of thinking that creates mindsets.

Association—See Free-association and Guided-association.

BALANCE—nemonik that prompts the mind to maximize success by maintaining the Yin-Yang balances, which prevents counterproductive rebalances.

Brainware—set of material components comprising the Central Nerve System. See Mindware.

Cognitive dissonance—individual rejection of correct information in order to protect incorrect subconscious information by evoking individual stress, fear, frustration, anger, aggression, etc.

Cognitive procedures—include memory and thinking.

Collective mindmode—rational way of thinking that deals with the rules of people.

COLLECTIVE—nemonik that prompts the conscious mind to maximize success by using Rational Thinking in order to deal with people.

CONCEAL—nemonik that prompts the mind to maximize success by hiding true information from the reality.

Concentrated-Relaxation—tool of Mind Management that evokes semiconscious dominance by reducing conscious awareness without falling asleep. It includes: muscle and breathing awareness; evoking semiconscious dominance; progressive muscle relaxation and breathing; deepening semiconscious; creating mindsets with mantras; returning conscious dominance; and mind-slide.

Conscious dominance—healthy mental state of being awake that is mutually exclusive from subconscious dominance and is fostered by concentration. See Mind.

Conscious—relatively small part of the mind that is dominant when awake; aware of the reality and the Self; manages the mind; and maximizes success with Rational Thinking. See Mind.

Conventional Thinking—educational competitive-memory based way of conscious thinking that aims to win competitions in the external reality by using free-association. See Nemonik Thinking.

Creative mindmode—affectorial way of thinking that deals with the future by generating mindsets that produce affecters such: as ideas, innovations, and intuitions.

CREATIVE—nemonik that prompts the subconscious mind to maximize success by using Affectorial Thinking in order to deal with unique problems.

DISPOSE—nemonik that prompts the mind to maximize success by decreasing the amount of matter that is under control.

Dynamic Thinking—part of Nemonik Thinking that maximizes external success by using the Nemonik Associative Network and guided-association. See Static Thinking.

Educational thinking—See Conventional thinking.

EFFICIENT—nemonik that prompts the mind to maximize success with a minimum of effort and resources by aligning with restoring forces.

External reality—surrounds your mind and includes space, matter, and time. See Internal Reality.

Firewall—cognitive tool that protects the genius from external and internal interference and manifests itself as cognitive dissonance and groupthink.

FOCUS—nemonik that prompts the mind to concentrate on maximizing success. Success is to obtain what you seek and escape what you suffer.

Free-association—cognitive procedure that activates spontaneously a link between two or more pieces of information. See Guided-association.

Genius—part of the subconscious that prevent a conscious information-overload by storing and processing that information and executing all inherited and learned skills.

Groupthink—collective way of thinking that is based on cognitive dissonance and evokes collective fear, frustration, anger, and aggression.

Guided-association—cognitive procedure that activates a memorized link between a nemonik and the icon, definition, potential questions, SWOT analyses, mantras, and proverbs for that nemonik. See Free-association.

Hegel, George Wilhelm Frederich (1770-1831)—German philosopher who introduced Hegel's dialectic describing the progress of knowledge.

Hegel's dialectic—during the first stage of this dialectic, a tested description of reality is introduced that is called the thesis. During the second stage, that thesis is challenged by a tested and conflicting description of reality that is called the antithesis. During the third stage, this conflict is resolved by merging the thesis and antithesis in a synthesis. The synthesis becomes the new thesis and the process repeats itself. Consequently, the development of knowledge is a process and, therefore, knowledge is temporary.

Humphrey, Albert, S. (1926-2005)—introduced the concept of SWOT-analyses.

Ignorance—lack of knowledge.

Internal reality—mental perception of the external reality. See External reality.

Knowledge—description of reality.

Lateral thinking—way of creative thinking that reformulates problems and looks at them from different perspectives in order to generate ideas and find solutions. Edward de Bono introduced this way of thinking in his book 'Lateral Thinking' (1970).

Manmade Extinction Problems (MEP)—climate change, nuclear war, dwindling resources, overpopulation, industrial pollution, extinction of species, and the overall destruction of nature that are caused by systemic cognitive failure.

Mantra—tool of Mind Management comprising a word or phrase that is repeated in the mind during semiconscious dominance in order to manage the subconscious genius. See Concentrated-Relaxation.

Memory—cognitive procedure that stores, maintains, and recalls knowledge in order to preserve it across space and time. See Thinking.

Mental Firewall—See Firewall.

Mental state—distinct level of awareness such as conscious dominance, subconscious dominance, semiconscious dominance, and unconsciousness.

Meta-thinking—thinking about thinking.

Mind Management—part of Nemonik Thinking that maximizes internal success by using Concentrated-Relaxation and Mantras. The aim of Mind Management is fostering conscious management of the subconscious genius by opening their semiconscious communication channel.

Mindmode—special way of thinking that deals with a particular topic. Mindmodes are Objective, Collective, Creative, and Reactive.

Mind—nonmaterial part of a person that includes all conscious, semiconscious, and subconscious processes and structures resulting from interactions between the brainware and mindware of that person.

Mindset—habituated set of rules or algorithms that generates affecters.

Mindware—set of nonmaterial procedures that could be compared to the software of a computer. See Brainware.

Mnemonic—cognitive tool that activates the memory. See Nemonik..

Nemonik Accelerator—cognitive tool that increases the speed of your thinking by fostering agreement during disagreement, while fostering disagreement during agreement. The nemonik-accelerator is based on Hegel's dialectic.

Nemonik Associative Network (NAN)—subconscious construction linking an exhaustive set of 21 keywords called nemoniks with guided-association to an almost infinite amount of well organized problem-solving tools.

Nemonik Thinking—conscious, semiconscious, and subconscious way of thinking that maximizes external success with Dynamic Thinking and internal success with Mind Management. Nemonik Thinking is the framework for the multitude of previously proposed ways of thinking including: analytical, collective, creative, critical, intuitive-emotive, lateral, left-brain, legalistic, logical, objective, rational, reactive, right-brain, scientific, strategic, and systems thinking.

Nemonik—memory and thinking tool that recalls associated information from memory and activates the processing of information in order to create new information. The nemoniks are: Focus, Balance, Efficient, Proactive; Objective, Collective, Creative, Reactive; Advance, Stay, Retreat; Accumulate, Preserve, Dispose; Act, Wait, Prepare, Accept, Reject, Reveal, and Conceal.

Objective mindmode—rational way of thinking that deals with the laws of nature.

OBJECTIVE—nemonik that prompts the conscious mind to maximize success by using Rational Thinking in order to deal with nature.

PREPARE—nemonik that prompts the mind to maximize success by getting ready for the most productive action.

PRESERVE—nemonik that prompts the mind to maximize success by maintaining the same amount of matter that is under control.

PROACTIVE—nemonik that prompts the mind to maximize success by restoring the balance before the problem grows.

Problem—anything that inhibits the maximization of your success. Many phenomena only appear to be problems. Therefore, the ownership, solvability, and priority of a problem have to be examined before you try solving that problem. Many problems are like monkeys. They sit on the shoulders of their owners and jump on the shoulders of the unaware. That happens especially to inexperienced managers and caretakers. Their charges complain to them about their monkeys and then the monkeys jump if given the chance. Delegating responsibility and authority will prevent a monkey zoo on your shoulders.

Procrastination—counterproductive delay. It is a thief of time and opportunities.

Rational Thinking—conscious way of thinking that subjects valid and reliable descriptions of reality to logic and reason in order to create new valid and reliable descriptions of reality.

Reactive mindmode—affectorial way of thinking that deals with the past by generating mindsets that produce affecters such as: skills, emotions, and feelings.

REACTIVE—nemonik that prompts the subconscious mind to maximize success by using Affectorial Thinking in order to deal with repetitive problems.

Reality—includes External and Internal realities.

REJECT—nemonik that prompts the mind to maximize success by rejecting the incoming information as a true description of the reality.

RETREAT—nemonik that prompts the mind to maximize success by increasing the distance to the goal.

REVEAL—nemonik that prompts the mind to maximize success by projecting true information to the reality.

Semiconscious dominance—drowsy healthy state of mind between conscious and subconscious dominance that is fostered by relaxation. Every time you fall asleep or wake-up slowly, you will slide through semiconscious dominance. See Mind.

Semiconscious—part of the mind that manages the communication between the conscious and subconscious through an adjustable firewall. See Mind.

Silent mind—mental state that activates the subconscious without conscious interference, which is often used to cope with emergencies. Also referred to as the Empty Mind or Zone.

Static thinking—way of thinking that aims to maximize winning in the external reality by adjusting the situation to a fixed way of thinking. See Dynamic Thinking.

STAY—nemonik that prompts the mind to maximize success by maintaining the same distance to the goal.

Subconscious dominance—healthy state of being asleep that is mutually exclusive from conscious dominance and fostered by relaxation. Nevertheless, the subconscious could become dominant during an emergency. See Mind.

Subconscious—largest part of the mind that is dominant when asleep; stores and processes bulk information; consciously inaccessible; and maximizes success with Affectorial Thinking. See Mind.

Success—obtain what you seek and escape what you suffer (Lao Zi [1, p. §62]).

SWOT-analysis—problem-solving tool that evaluates the Strengths, Weaknesses, Opportunities, and Threats of each nemonik in comparison to the actual situation. Strength is an intrinsic advantage; Weakness is an intrinsic disadvantage; Opportunity is a potential advantage; and Threat is a potential disadvantage. SWOT-analyses were introduced by the American business consultant Albert S. Humphrey (1926-2005).

Systemic Cognitive Failure (SCF)—static thinking in a hyperdynamic reality that has created a triangle of doom. Education fosters science; Science fosters technology; Technology fosters manmade extinction problems. Systemic cognitive failure is maintained by cognitive dissonance and groupthink.

Thinking—cognitive procedure that processes available knowledge in order to generate new knowledge. The main ways of thinking are Static, Conventional and Dynamic thinking. See Memory.

Triangle of Doom—education fosters science; science fosters technology; technology fosters manmade extinction problems.

Unconscious—unhealthy mental state that is characterized by the absence of mental activity.

WAIT—nemonik that prompts the mind to maximize success by delaying action until it is the most productive time for that action.

Winning—defeating others in competition.

Wisdom—productive application of knowledge.

Index

List of Figures

List of Tables

Made in United States
North Haven, CT
05 November 2026

21758667R60201